Arctic Screaming:

Journey to the Front Line

of the Climate Crisis

Second Edition

Chad Kister

Arctic Screaming

Printed and Distributed by

Zip Publishing and Progressive Publishing

Columbus, Ohio

ISBN: 978-1-934849-95-8

Acknowledgments

Thanks to my mother Joanna Kister, my father Robert Kister, my brother Scott Kister, his wife, Stella and my nieces, Zoe, Kara and Mia. Thanks to my grandmother, Bernice Hunsinger and my late grandfather Carl Hunsinger.

Thanks to those who have helped me to continue my writing and public speaking: Harvard Ayers, John Cooper, the Mast General Store, Hanes Boren, Footsloggers, Dick and Eleanor Edgar, Art and Peggy Gish, Ann Bonner, Keith McPencow and Pat Westfall.

Thanks to Deborah Williams, Guntner Weller, Josefino Comiso, Glenn Juday, Bruce Molnia, Pam Miller, Fran Mauer and all the others who took the time for me to interview them.

Thanks to Sarah Ortman and Ted Bernard. Thanks to Ralph Izard, Dru Evarts, Tom O'Grady and many more.

Thanks to the Alaska Coalition (www.alaskacoalition.org), the Alaska Wilderness League (www.alaskawild.org), the Northern Alaska Environmental Center (northern.org) and the many other organizations working to protect public lands in Alaska.

Thanks to Greenpeace, the Sierra Club and the many other organizations working to combat climate change and implement solutions to our suicidal path of no return.

Dedication

This book is dedicated to the thousands of activists and organizations who are working to reduce greenhouse gas emissions at all levels. Efforts from the local to the regional, state, national and global are critical to tackle what is by far the greatest threat that we face.

A further dedication goes to those who will be inspired through the words of top climate scientists and native peoples to join the rising environmental movement and work to reduce and ultimately stop greenhouse gas emissions while creating a nicer, cleaner, healthier world and sustainable economy.

Through a change to mass transit, efficiency, wind, solar, conservation, muscle power (walking and bicycling, etc.) and other renewable and ecological sources of energy, we can meet our needs while improving our health and safeguarding the future. It is the moral duty of our day to dedicate time to learn about this critical issue, and make the changes before it is too late.

Contents

Arctic Screaming

Preface

Boating the Arctic Ocean with Inupiat guide Robert Thompson, where all was ice a decade prior, the scale of the change in such a short time foretold global catastrophe. Flying from Arctic Village to Fairbanks, the forest was scorched for miles, with fires blazing in all directions. Our planet is burning. Smoke was so thick as we approached the airport that our pilot said (later) that it was the most challenging landing he had ever had in his 30-year career.

Major changes are wreaking havoc on the Arctic environment, melting the ice of the Arctic Ocean, melting the permafrost, killing millions of acres of forest, eroding shorelines, crumbling native villages into the sea and disrupting Arctic ecosystems. The extinction of Arctic megafauna like polar bear, walrus and seal is predicted, unless massive changes are made.

While the impacts are being seen most dramatically now in the Arctic, it is not just the polar bear that is threatened. Well respected scientists warn that billions of people could die this century from climate change-caused storm events, heat waves, flooding, drought and disease.

Predictions of when we will see an ice-free Arctic Ocean changed from 2080 in 2005 to 2040 in 2006 to as early as 2012 in December, 2007, after scientists realized the implications of the record 2007 melt. Warming is accelerating so fast, that scientists fear the tipping point to unstoppable climate change they have long feared may have been reached as the warming is triggering positive feedbacks that cause more warming.

I experienced the massive, horrific changes to the Arctic in two expeditions in 2004 and 2005. In 2004, I journeyed across the continent by train, ending in British Columbia. I took a ferry to southeast Alaska where I explored the Tongass National Forest by foot, kayak, bus, ferry, Greenpeace boat and Greenpeace helicopter. Next, I went up to the Arctic to visit with the Gwich'in in Old Crow with a group from Appalachian State College. I attended the Gwich'in Nation General Assembly in Old Crow.

Next, we boated hundreds of miles down the Porcupine River from Canada into Alaska with Gwich'in guides. We then took a bush plane into the Arctic National Wildlife Refuge. The tundra was much drier and the landscape far different in just the 13 years since I had backpacked and rafted hundreds of miles through the Arctic in 1991. Giant sinkholes caved into the permafrost, riverbanks eroded into the sea and foot-wide cracks opened along lake banks, where the permafrost was melting.

The climate changes are being caused mostly by the burning of fossil fuels. While some think it unfathomable that we can change the climate, we need only look at a satellite photo of the Earth at night to realize the scale to which we have conquered the planet, and are burning fossil fuels. Carbon dioxide levels are up 37 percent and methane levels have gone up two and a half times in the atmosphere since 1850. Both are greenhouse gases, trapping in heat like the panes of a greenhouse.

Fossil fuels should be renamed carbon reserves to better describe the true purpose for coal in particular and also oil and natural gas. As long as we leave them in the ground, and switch our energy sources very fast to proven renewables, efficiency and conservation, we can avoid the worst of the climate crisis.

As we learn more about tipping points, and the attractor states of the climate, we should be very worried. Scientists warn that we may be on the very rapid shift to a super-warm earth that has not been seen in 55 million years. While we should hope they are wrong, the increasing scientific support of this argument and the severity of the implications means we should devote commensurate resources to understanding and acting on this issue.

Instead of leaving it at the doom and gloom of our current path, this book also shows how we can make massive reductions and ultimately eliminate our emissions of greenhouse gases through increasing efficiency in every sector of our economy, and changing to renewable energy sources. This can be done using existing technology and economic resources.

Rather than sitting back in despair and giving up on our planet and all life to come, we must take action like never before. We have seen amazing changes in industry and lifestyles when the U.S.

mobilized for World War II. World War III is here, but it is a different kind of war. We are collectively at war against our life support system, and we must make tremendous changes very fast if we are to leave a livable planet for future generations.

We not only can, we must make the shift to conservation, efficiency, solar, wind, geothermal, waste methane, wave and tidal energy if we have a chance of leaving a planet with the semblance of the diversity of life as we know it to our children. The facts are far too proven by thousands of scientists to ignore. The impacts are far too severe to leave our fate to politicians guided by the interests of the fossil fuel industries that are causing the destruction. We must commit ourselves to being citizen activists, and make a difference in the world instead of being a bystander as our beloved planet burns.

We all have the power to make a difference to slow climate change. Through this climate journey across the continent and far into the Arctic and Arctic Ocean, you will understand the latest findings and the credibility behind them so you can then make the information widely known to family, friends, media and decision-makers. Such a vast amount of understanding is unfolding at such a breathtaking pace that it can be quite overwhelming. I have worked to take the most powerful and credible of it, and put it into one book, along with my first-hand witness of the massive impacts already seen.

Armed with the facts, make it a routine to contact your local, state, and federal representatives, despite how they vote, and educate them about the urgency of this crisis, and your support for strong mandatory measures to cut greenhouse gas emissions. Also make sure to contact the media with letters to the editor, calls to talk radio and more.

The latest findings show that we all should be very alarmed, and engaged as full time as we can in making critical changes. One of the best is passage of cap and trade carbon reduction legislation. This would use the free market to find the most cost-effective means to cut emissions. Then we can work on the strengthening of the legislation until emissions are cut at least 80 percent.

Meanwhile, we should begin efforts to increase long-term carbon sequestration through sustainable means, such as old growth forest protection and restoration, and building up carbon in soils with composting and organic agriculture, possibly enhanced by burying charcoal. The idea of capturing carbon from coal power plants is unwise: the key is to stop releasing that stored carbon in the first place, by stopping the burning of coal, oil and natural gas and instead generate our energy from renewable sources.

The scariest thing that I have encountered in speaking around the world on this issue is people giving up: people who say it is not even worth it to try to do something about climate change. While the facts are dire, and getting worse every day with the increasing emissions and understanding of the forces we are unleashing by them, we cannot give up. We cannot give up on future generations. We cannot give up on all life on our planet, and all life to come. We must get involved politically and make a difference, before it is too late.

Arctic Screaming

Chapter 1

Anthropocene: On a Runaway Train to a Hellish Climate

Think of our planet like a freight train, accelerating along its tracks on a high mountaintop where life flourishes. The top is a near equilibrium of amazing abundance and stability that life has evolved to fit into for thousands of years. But downhill on the tracks, lays an inhospitable desert with heat so unbearable as to wither nearly all life to death. For thousands of years, the train has remained on the top, moving very slowly back and forth, with the climate relatively stable.

The mountaintop is the only place full of life in its grand abundance and diversity, albeit heavily trampled by past environmental abuses. While very degraded, it still has most of its constituent parts, and above all the climate as we know it: critical to the survival of the great diversity of life today, including us. It is only up on the mountaintop, with that relatively stable interglacial climate, that creates the vast array of life that we have today.

On the top, our planet is easily repairable into its far grander former abundance when hundreds of millions of salmon returned to the rivers of California, Oregon and Washington States to feed millions of Native Americans, and tens of millions of buffalo roamed the Great Plains.

But on down the tracks, all hope of restoration, or preserving the remaining wonders of our planet, are beyond hope. Keeping our 6.75 billion people alive will be near impossible in the climate down the mountainside. The loss of stable rainfall for crops, rivers overflowing their banks, sea levels rising, increased frequency of hurricanes and tornadoes, and tropical diseases in temperate regions threaten massive loss of life.

Realize the seriousness of this issue. We are not talking about remote speculation. This is the undeniable course that we are on: just look at the Arctic sea ice, and all the documentation in this book. The permanence is eerie. When working to save 3,000 year-old redwood and sequoia trees, the age of the giants is impressive. Now we are talking about shifting to a climate not seen for 55 million years, threatening to wipe out the vast majority of species alive today, including those that have survived tens of millions of years, and threatening the survival and quality of life of most of our own species. That is the train we are on, whether we want to deny it or not, and we had better wake up, because we are the ones fueling its motion down toward a hellish oblivion of mass suffering, chaos, disease and death.

Our vision has been clouded by the wonders and might of our civilization's ever growing technological power that makes us feel impervious to all threats. That fuel for the train is fossil fuel consumption and deforestation. Those most knowledgeable of the climate situation of today beg people to take heed of the gravity of the situation we face, and realize how much we depend upon natural systems to provide our basic needs of clean air with oxygen, clean water, and food. These are the systems that are collapsing in the climate crisis.

The fact that we have increased carbon dioxide levels 37 percent and methane 150 percent in our atmosphere puts any thought that we are an insignificant influence to rest. The science is clear that it is our activities that have caused the changes. Had we listened to the credible warnings of two decades ago, we could have made the adequate investment to switch to a fossil-free infrastructure that also solves so many other problems, and greatly improves our health by nearly eliminating pollution. We likely could have stabilized carbon levels, and begun to reduce them to 350 parts per million carbon dioxide (the amount in 1987) as top scientists say is necessary.

Instead, we have increased carbon dioxide emissions a record 2.9 percent in 2007 and 3.5 percent in 2008, more than three times the rate of increase of the 1990s. The reason is the power and corruption of the world's political systems by the coal, oil and auto industries, which have invested their fortunes into coal and oil reserves, and gas-guzzler futures. Their investments would be near worthless if we could meet our energy needs through renewable power and travel primarily by fuel-efficient train.

In the beginning half of the twentieth century, industries abused their power by buying up tracks and turning them into roads so that the oil, auto and tire companies could increase profits. This hastened the demise of passenger rail as the primary mode of transportation. Next, oil companies bought patents for fuel efficient cars that were then kept off the market and both oil and coal companies bought patents for solar panels and wind generators to keep them from production. Now these industries are engaged in multi-billion dollar campaigns against efforts to tackle climate change and shift to renewable energy.

Coal and oil industries have massively pushed the freight train Earth toward the precipitous edge of the mountainside, at an accelerating pace. Instead of investing their billions of dollars into renewable energy and energy efficiency improvements, the fossil industries are orchestrating the release of vast amounts of carbon stored for hundreds of millions of years in our ever expanding coal mines and oil and gas fields.

Chances of stopping this train from roaring down the mountain into a hellish climate grow increasingly lower every second of every minute of every hour of every critical day in these pivotal times. The technological might that has moved mountains for coal and highways is all based on fossil fuels that are running out, and that we should not be burning. It would be near impossible to push freight train Earth back up the mountain if we let it go down.

The slope in this metaphor is caused by what scientists call positive feedbacks. This is where the warming we have initiated triggers natural mechanisms that cause more warming, a runaway greenhouse effect until a new stable climatic system is reached. The problem is that next stable condition is about 20 degrees Fahrenheit warmer, last seen 55 million years ago, when tropical conditions existed in the Arctic and Antarctic region. We are unleashing great forces of nature that are just now being understood, that can force our climate into inhospitable levels not just for the 60 percent of the planet's species already known to be threatened with extinction this century, but for humanity as well.

The scale of the melting ice in the Arctic to alter Earth's temperature exceeds that of all the fossil fuels we have already burned. The permafrost contains twice as much. Then there are the methane hydrates, that scientists believe were the largest factor in reaching the super-hot conditions seen 55 million years ago, when most of the life on our planet was wiped out. This is frozen methane that has been stored thousands of feet deep on the ocean floor throughout the world. It contains 3,000 times the amount of methane in the atmosphere today, and there are signs it is beginning to melt, releasing bursts of methane that bubble to the surface.

The Arctic sea ice is like a switch that rockets the planetary train to overdrive when melted. With ice conditions year round in almost all of the Arctic, as was seen until recent years, it acted like a brake on the train, reflecting much of the sun's energy, giving us a fighting chance to stop before we go over the edge. Once sea ice melts, warming conditions threaten to rapidly melt permafrost, and possibly even methane hydrates, catapulting the planet to super-hot, deadly conditions. When a house is on fire, you get outside. What do you do when the whole planet is burning?

The horrific conditions of the destination into which we are headed is the most troubling, and something few scientists realized until very recently. Several years ago, many would have described my analogy in terms of a flat surface, that once we get that climate train moving, it can be hard to stop, and could cause horrendous damage with a few degrees temperature rise. But few understood that once we get it moving past a certain threshold, the slope takes over and even if we stop the engine, it is going down toward the unbearable conditions none of us would want to leave to future generations.

We take for granted the life support system in nature that keeps humanity alive far too much. We are talking about the life or death of billions of people, not to mention millions of other species. Such is the value of the climate on the mountaintop for our beloved Earth: the critical element that brings life-giving rains at just the right amount and time, and temperature ranges in which all plants and animals are evolved to fit.

Climate change has turned from a few degrees warming about which we should be concerned, to a likely terrifying lurch into such temperature extremes that the very survival of the majority of humanity this century is being put into question by top scientists like Sir James Lovelock.

We are triggering so many positive feedbacks of such a gigantic scale, that the runaway greenhouse effect that we are unleashing could exceed the worst case IPCC predictions of 6.4 degrees Celsius, which are already horrific. Our planet is in critical condition, the situation very dire. But, an optimist at heart, as all activists inherently are or they would just sit on the sidelines, I will never give up hope.

To give up the hope that we can make a difference, since it is our species that has caused the crisis, is the ultimate nail in the coffin of the welfare of all life on our planet for all foreseeable time to come. Through the collective ingenuity and the might of humanity, we need to mobilize for the enormous needed changes. For now, the best thing to do is speak up persistently, and demand strong action from government and industry leaders, while trying to reduce our own carbon footprint.

All of the effects detailed in this book, as horrific as they are, are bound to get much worse very fast. Even with no new emissions, our planet will continue to warm for another 35 years as there is a thermal inertia with the length of time that it takes for the oceans to absorb heat. Also, through the global dimming caused by all the soot pollution in the atmosphere, we are about two degrees cooler than we otherwise would be. Ironically, if we were to immediately stop burning fossil fuel and wood fires, temperatures would immediately rise in just a few days. But we surely can't keep burning fossil fuels, either, or on down the tracks to a deadly climate we go, with no way back.

The 220 billion tons of greenhouse gases we have already emitted have moved the train to where the slope is beginning to take over. Meanwhile, we are fueling the locomotive with three times the global emissions of a decade ago, and the slope increases as we go down the tracks.

Drastic change needs to be made very fast, and we need more people demanding faster action from lawmakers and industry. Through a broad understanding of the facts presented in this and similar books by the citizenry of our planet, we can work to collectively realize that the issue is not our lifestyle, it is our survival, and particularly the survival of future generations that is at stake.

Working with organizations such as Greenpeace, I began asking for action on climate change in 1989. Then, few understood the threat global warming presented. It was after working for a year with Ozone Action, which merged with Greenpeace USA in 2000, that I realized how serious the issue was.

I spent six weeks in The Hague, Netherlands at the United Nations Conference on Climate Change, working with Greenpeace to show support from the U.S. for strong mandatory cuts in emissions. I was impressed with Greenpeace, as they combined banner drops with suit and tie negotiations very effectively and professionally. Greenpeace activists chained themselves to a coal barge, shutting down Rotterdam Harbor. I witnessed Greenpeace activists chain themselves to the truck they stopped in the middle of the road in front of the UN building. They took the tires off and poured cement onto the chained-in volunteers. Meanwhile a siren blared: an effort to sound the alarm on this issue that we should have made number one back then.

I cheered support for the brave souls willing to risk death to get the message out on this critical issue. I also passed out fliers explaining the purpose of the demonstration, along with dozens of other Greenpeace volunteers. One of the facts that startled me the most was how 42 percent of the polar ice had melted in the prior 40 years. Having trekked through the Arctic Wilderness twice, I began looking into it in much greater depth.

Two years prior, Greenpeace sent researchers up to the Arctic to get witnesses to the change that scientists knew was occurring, and publicized those findings. Greenpeace "wanted to work to educate Americans on Global Warming and raise the level of concern," said Dan Ritzman, who worked on the project. "The second component was we wanted to do something directly to combat climate change. Greenpeace was unique in that it is international, and they did the science to prove that we are getting to the critical point of no return. We cannot burn a quarter of our known reserves. So there is no point in finding new reserves. We should use that effort to find renewable energy sources."

Chapter 2

A Climate Journey to Alaska by Train and Boat

With my constant desire to minimize my own carbon footprint, my third journey to Alaska, and the Arctic, began by train. Sacrificing safety and comfort, I drove four hours through the night to catch the 5 a.m. train in Toledo, a regular chore to get from my rural southern Ohio home to the luxury and fuel efficiency of passenger rail. The trains only come through in the early morning hours, with the terrible train system that we have in America today.

I unloaded my gear and found a window seat, exhausted from the ordeal of packing for a three-week speaking tour, and an entire summer in the far north on two expeditions, one in southeast Alaska and one in the Canadian and Alaskan Arctic, then driving through the night with no sleep. Such a relief and feeling of utter joy and fulfillment it was to lounge in my seat, with the clicketty-clack of the rail: a soothing sound. Without the stress of driving, when one split second mistake can mean death, train travel is so much more relaxing.

But looking at the landscape, I knew all was not well with the world. Large power plants were lit in eerie scenes of industrial blight at the shore of Lake Erie and Lake Michigan. While we should be rapidly shutting down power plants, replacing the energy with conservation, efficiency and renewable energy, we are actually building more of them, with several proposed just upwind of my rural home. Scores more are proposed nationwide. Such is the industrial scene all throughout our continent, on lakes and rivers where otherwise spectacular scenery would dwell.

With ecological eyes, I watched mile upon mile of corn, wheat and soybean fields where previously had stood the largest contiguous forest in the world, with trees 500+ years old. I had just spent years battling Ohio Valley Coal Company and its owner, Robert Murray from mining under Dysart Woods, among the last of the .004 percent of old growth forest left in Ohio. Now I was on a journey that included research for a book and future speaking tour to protect the largest old-growth forest left in the U.S.: The Tongass National Forest in Southeast Alaska. Along with being critical for the salmon that form the base of that ecosystem, logging is a climate change issue. Deforestation causes about 30 percent of greenhouse gas emissions.

In Chicago, I changed trains to the California Zephyr, heading over the Rocky Mountains for an intense three-week speaking tour of California, including time to visit my brother and his wife in nearby San Francisco. By train, one can really see the scale at which we have altered the landscape, if one knows how the land was like centuries ago. All of the logging, tilling and bulldozing has released vast amounts of carbon, and continues to release carbon that had been stored in the soils, making them less fertile and less able to sequester carbon with forests in the future.

In Denver, I picked up a growler of India Pale Ale at the Wynkoop microbrewery across from the train station: Denver's first brew pub. Heading up the mountains from Denver, we were delayed for hours by mile-long coal trains that came lumbering in from Utah and Wyoming. About a hundred train cars, each filled with more than 120 tons of coal rolled by as we waited on the side. Coal that has been in the ground for hundreds of millions of years gets priority over people in our country: the perverted priorities of coal's criminal corruption in politics. In Ohio, we spend hundreds of millions of taxpayer dollars to build rail lines for coal mines, but have among the worst passenger rail systems in the country.

Up in the mountains, we watched rafters cascade down the white-water headwaters of the Colorado River that the train went by. Mountain streams and rivers, and all of the animals and people that depend upon that water downstream are threatened by climate change. As the snow packs and glaciers that once fed these streams year-round dry up, so do the streams. We finally reached western California, and the smog and endless development of the Bay area.

The train went over massive highways, and I had a wonderful feeling of elation as I lounged in my seat, closing my eyes to rest as desired, and skipping the chore and stress of driving in such mayhem. Mass transit can not only exponentially slash our carbon emissions, it is also a much better quality of life, and hundreds of times safer than driving. We forget about the trillions of dollars we have invested in roads, highways, cars, insurance and gas when comparing the paltry billion or two we give Amtrak each year.

The 12-lane highways of San Francisco and Los Angeles were a stark reminder of the massive burning of fossil fuels that is accelerating on our planet: causing about half of greenhouse gas emissions. Even in Yosemite, the smog drifts in, threatening the giant trees.

Hiking in Yosemite, I was reminded of my first journey to the Arctic, where I backpacked and rafted hundreds of miles, from Prudhoe Bay into and throughout the Arctic Refuge. I had conditioned in Yosemite. I had barely survived my first Arctic Quest, nearly dying of starvation, then drowning and hypothermia.

Now I understood the complexities of the greater ecosystems around me far greater because of my extensive travels and studies. I knew how rare the amazing wonders of Alaska are, and how fragile. The collapse of salmon fisheries along the west coast of the lower 48 foretell the fate of the salmon runs in southeast Alaska, unless the forests are protected.

In one of my many past trips by train to California, I took a ferry to Santa Cruz Island, part of the Channel Islands National Park, and backpacked around it. On the way back, we saw a pod of gray whales mating. After this journey around California, I would meet gray whales again in Alaskan waters, though by a much longer route, crisscrossing the continent by train.

Arctic Screaming

Millions of native Americans once lived on the hundreds of millions of salmon that once swam up the rivers and streams each year from Washington to California. Now, most are wiped out, or in the tens of thousands of fish at best, teetering on the brink of survival.

The old growth forest was destroyed. Without the roots to hold the land in place, landslides blocked streams, and massive amounts of silt washed into the streams. Salmon eggs need clean streams with gravel bottoms to spawn.

In addition to the logging, dams blocked migration routes. Development further damaged the salmon with silt and alteration of the riparian habitat. Now these fish are under assault from climate change as well. Having traveled all throughout the Pacific Northwest, where I heard of the plight of the salmon, I could not wait to spend some time in southeast Alaska. The Tongass National forest there is one of the few left on Earth that still have salmon in their original numbers: tens of millions of five species of salmon.

I still needed more gear for my expeditions in southeast Alaska and the Canadian-Alaskan Arctic. I picked up a water bottle at the Yosemite store. Later, a few blocks from the hostel where I stayed in downtown San Diego, I found a used army store and a $30 tent. While not exactly Arctic gear, I figured I would be there in the summer, and with climate change, it should do, assuming everything worked out perfectly and several thousand dollars in credit card debt.

I returned home on the Southwest Chief, from LA to Chicago, then on to Toledo, and Columbus for my 15th high school reunion. Exhausted from dancing past 2 a.m. after the reunion, the next day I drove to Cleveland, to catch the train headed east. Driving in Ohio, I could smell the mega-farms from miles away, as the local residents must do all the time.

Much of the state now stinks from massive megafarms where cattle, pigs and chickens are grown in abysmal living conditions. Livestock production was recently found to contribute as much as 18 percent of greenhouse gases, mostly from methane. The nitrogen fertilizer used on the grains to feed the livestock also contributes a significant amount of emissions. I hopped on the train in Cleveland with an entirely new set of gear, going through Buffalo, New York, then Toronto and on across Canada toward British Columbia.

It would have been much easier to go up through Windsor to get to Toronto, headed west and north instead of way east and up through Buffalo then back west. But excessive security delays by paranoid police after September 11, 2001, along with other factors put a stop to the train from Chicago to Toronto in 2004. I wonder how many more people will be killed by the auto accidents and increased greenhouse gas emissions as people drove the route instead of taking the train.

Such is the lunacy of police state USA, which under the former George Bush Regime has been the primary cause of global inaction on climate change. While we increasingly invade the privacy of travelers in America, and have trampled on the Constitutional rights of all citizens with the Patriot Act, the failed former George Bush regime fanned the flames, censored the science and stymied action to prevent the mass murder of billions of people this century through climate change. Obama is a refreshing change in priorities.

The previous September, I attended a lobby week in Washington D.C., sponsored by the Alaska Wilderness League, in which scores of citizens from around the country came to our capital and met with Representatives and Senators asking that they protect the Arctic National Wildlife Refuge as wilderness. While I was there, Appalachian State College Anthropology Professor Harvard Ayers asked if I would be interested in guiding a group of students up to the Arctic. I jumped at the opportunity.

On the way, I would spend an intensive month in southeast Alaska, for a book about the ancient forest and the vibrant ecosystem and peoples that it supports. Then I would meet the group in Whitehorse, traveling all by land and water to that point except for two amazing flights through the Tongass National Forest and the clear-cut devastation by Greenpeace helicopter, from their boat Arctic Sunrise.

In Alberta, the train followed the Athabasca River, with snow-capped mountains towering above. The train had a dome car, allowing passengers to look up at the mountaintops and sunsets. But just downriver, the river was being decimated by the single most environmentally destructive project in the world. A wetland area the size of Florida is being strip mined. The tar sands are then cooked using vast amounts of fossil fuels, and dirty oil is extracted to be burned in the U.S.

Canada's exploitation of tar sands for oil is producing horrendous amounts of greenhouse gas emissions as well as massive destruction. Largely because of these emissions, in the spring of 2006, Prime Minister Stephen Harper announced that Canada did not intend to meet the greenhouse gas reduction targets that the country agreed to in signing the Kyoto treaty.

In the thousands of miles on the train through the U.S. and Canada, I witnessed the landscape, wondering what it had looked like 300-years ago. Seeing that the vast old-growth forests had been felled, the bison decimated and the Great Plains near entirely dug up, replaced with suburban sprawl, strip mines and chemical agriculture, the scale of the change that we have made to the landscape is incredible. The fact that it is changing the chemistry of the atmosphere, and thus the climate is logical when you see the amount of coal that we burn every day. It is simply undeniable in today's scientific understanding.

Chapter 3

Northern Forests Hammered by Climate Change and Logging

After a spectacular ride, the train ended at Prince Rupert, British Columbia, where I camped for several days near the ferry terminal. Logging forms the base of this community, and talk of it was everywhere in town. I wondered how many ancient trees were being lost every minute. I boarded the ferry to Alaska, selling one of my Arctic Quest books while in line. I needed every cent I could to buy food later in my journey.

The ferry ride was amazing, with mountains of forest rising from the ocean of the inside passage. But the scars of past clear-cuts were exceptionally visible, with tiny Christmas-tree sized trees growing where once there had been giants. In addition to the massive clear-cutting are the horrific impacts climate change is wreaking on the forests that we have struggled so hard to protect.

The yellow cedar, which can live to be a thousand years old, is dying. Paradoxically, its roots are freezing because warmer temperatures have caused a decrease in snowfall. The snowfall once buffered the roots of the cedar from the winter cold. With drastically lower snowfall levels, the roots lose this insulation, and die from the cold. On a much larger scale, climate change killed millions of acres of forest in Alaska through increased insect outbreaks, and the devastation is spreading worldwide.

Marko Scholze of Bristol University found that half of the world's major forests will be lost if global temperatures rise by an average of three degrees Celsius. Scientists predict this increase in temperature or more is likely this century, unless massive changes are made to reduce greenhouse gas emissions. The study found that there would likely be a loss of forests in Amazonia, Europe, Asia, Canada and Central America of up to 60 percent.

Those who think global warming might be a good thing, because they don't like the cold or the cost to heat their homes need to realize that cold temperatures safeguard the life support system that we all depend upon. When we lose it, we can lose the forests, which provide the oxygen for us to breathe, as well as clean water, clean air and so much more. And this is just one of the myriad of other impacts that we are just now realizing are caused by climate change, and they are nearly all very negative.

Throughout southern Alaska, vast stands of forest are dead. Climatologist Guntner Weller said climate change has caused large infestations of parasite disease species like the spruce bark beetle in southeast Alaska that have laid waste to unprecedented large areas of forest, destroying millions of acres of spruce. "That is undoubtedly caused by a warmer climate," the scientist said.

Tom Moran reported in the Fairbanks Daily News-Miner that the forests in Alaska have been thinning out. More heat but the same amount of available water is stressing trees, making them more susceptible to insects such as bark beetles. The Christian Science

Monitor reported on June 12, 2001 that scientists estimate that a third to a half of all the white spruce in Alaska had been killed in just 15 years.

On the Kenai Peninsula, more than four million acres of spruce forest have died because spruce bark beetles have attacked the global warming-weakened trees, federal scientists report. Timothy Egan wrote in the New York Times, "Century-old spruce trees stand silvered and cinnamon-colored as they bleed sap."

Dead forests reach all the way to Fairbanks, moving north.

"In Fairbanks and other places, the bugs normally go south (or hibernate) but now they stay all winter," said Arctic Village Elder Sarah James. "There are strange bugs like the beetles that are killing the spruce trees. The spruce beetles are mainly south of Fairbanks, but it will probably get to our village as it gets warmer," the elder predicted.

Rather than just small changes that we can adapt to, the effects of climate change seen in Alaska show that climate change is moving so fast and throws so many unpredicted punches as to cause devastating impacts for which we cannot adapt, said Alaska Conservation Solutions Director Deborah Williams.

"When most Americans think about global warming, they might imagine glaciers melting, and ice caps. That is something that has a clear cause and effect. The stories that have not been told well yet, in general but particularly in Alaska, are the increase in disease vectors and disease vectors that global warming enhances."

“The disease, insects and other secondary impacts of global warming are the scariest,” Williams said. She stood on a cliff, and nearly all the larch forest had died. “You just look at these huge dead stands of larch trees.”

“A story that has been covered better is the spruce bark beetle infestation in south-central Alaska. But the larch story hasn’t. Then you look at the spruce bud worm, and you see the tips of the magnificent white spruce curled up and diseased. And you look at the leaf minors, the aspen and birch and this is a devastated forest. Glenn Juday says that he doesn’t think if climate change continues that white spruce forests will be in existence in Alaska. They will be gone, gone. This whole forest ecosystem will be gone.”

The four degrees warming that Alaska has seen in the last 50 years has already triggered massive devastation, Juday said.

“Those warm temperatures are often the key environmental trigger for natural disturbance agents. In a cool boreal region, warm temperatures previously have been limiting. So the insects and their life history strategy were of low population background levels until that infrequent warm summer. Then that warmth is the signal and their populations go on a big upswing and they have big outbreaks and they attack and kill trees. That is a natural process that the boreal forest is well adapted to.”

“What we appear to be confronting is through climate warming that outbreak level of some of the insect agents that can kill trees is coming more frequently, more severely over greater areas and is becoming itself a factor that one could project would completely change forest types over large landscapes. And in fact we have an example of that in the Kenai Peninsula and the whole south-central Alaska spruce bark beetle system.

"Based on what we know, if we have climates similar to the scenarios that were produced, that if it were that much warmer, well short of the point of no growth, you could have stress mortality on a huge scale just from insect outbreaks that are triggered and finally just from acute stress to the trees.

The scenarios of warming that Juday studied a few years ago pale in comparison to today's predictions. "The bottom line, to encapsulate the whole thing, it appears that based on what we know in these scenario projections, that climate warming of that magnitude would be a major, major disruption of the entire boreal forest system. That disruption would amount to large-scale mortality of some of our more productive forests and the necessity for some tree migration and movement of trees into areas where they are currently poorly productive." The scientist noted there are "a lot of challenges in getting all of that to happen in a fairly short amount of time.

A little more than 4 million acres have been affected by the spruce bark beetle. The larger or overstory trees that form the main forest canopy have been subjected to mortality. That means that generally, the guys in the plane looking down on the ground can map the area as showing mortality from the spruce bark beetle infestation. That threshold before they can identify it from a plane is about twenty percent. When 20 percent of the overstory trees are dead from the spruce bark beetle they generally are able to recognize it from the aerial survey and say yeah, the trees down there are killed by bark beetles. And in this outbreak the level of mortality has gone from this lower threshold of 20 percent all the way to 100 percent.

"This outbreak is so severe that a very substantial fraction of the area that has been affected, more than 4 million acres, has 100 percent overstory mortality. Generally trees that are four inches in diameter or smaller are not killed. In those places, it may be simply that the overstory dies, then the understory starts to mature.

However in a lot of places, and a substantial portion of the area, there is not the understory of the spruce, and the spruce has died over such a huge extent that the seeds cannot glide from surviving trees far enough to reach these depopulated localities. Spruce cover has been eliminated from substantial areas.

"We don't really know what is going to happen on a lot of those sites. We don't know if there will be a colonization of spruce that will grow back and if they do, when they exceed the four-inch threshold and start to get bigger, will they get hammered too from the spruce bark beetle? As near as we can tell, these outbreaks track these warm temperature episodes. So the outlook is not too favorable for spruce persistence in localities in southcentral Alaska, around Kenai, Anchorage and the copper river valley.

"I just looked at some data and tree cores that have been taken down there that have been sent to me from many sites. What we can say is many of the overstory trees are in the 170-270 year-old range. So it is pretty clear that nothing on this scale or intensity has happened in at least a century and probably 300 years and based upon what we know of climate variability, if you go back 300 years from the present, you go back to the little ice age, so you have to go back a thousand years before you reach a point where climatically it was likely that anything of this scale has happened in the past.

"This is an extremely powerful event that has happened, and if the climate doesn't turn back cooler, it probably represents a semi-permanent change in the landscape by reducing the abundance and age structure of the spruce forest. And as to what will replace it, if any trees, we don't know and it does not look possible in places that there will be a good chance for long-persisting forest conditions where there used to be fairly productive spruce forest.

"Anything that is dependent upon the older spruce succession stages is going to be hugely reduced in abundance. And these include the larger-year-round resident woodpeckers, the black back and the northern tree-toed. This includes all the insects and lichens, all the organisms that are specialists in the older forest as well as organisms that may use older or mature forest cover for a certain time of year such as bears, and moose," Juday said.

"To the degree it is not available to them, they will likely experience some population reduction. There may be some trumping factors. Especially if there are widespread fires there could be a big increase in moose numbers. But already we know moose population density is so great on the Kenai Peninsula that they prevent the Kenai Peninsula birch from successfully growing up into trees in a lot of places. So with the spruce eliminated horribly, you would expect it would shift toward birch, but that may not be possible with the moose populations that are there. And that may be why we get stuck in this non-forest semi-permanent condition.

"It is a big shift in everything. The large woody debris and the logs that represent the eventual fate of all of the large trees down there end up in stream channels. They are extremely important as an aquatic habitat factor. There is going to be a big influx of that material over a period of time, and then it is going to be cut off because no matter what happens from now on, there is going to be a huge age class imbalance in the forest. A few decades down the road we are going to see some aquatic habitat impacts of this, with a big reduction of the input of new dead log material. It is just a complete chain of events that have been set off. It is not going to be a short-term effect. We are talking about landscape transformation for very long periods of time even if we make the assumption that everything would go back to normal tomorrow, which does not seem likely.

"It just so happens that most of the forest tree species, especially in the boreal forest region are simultaneously experiencing huge levels of outbreaks of insects that cause tree mortality. The best example would be the Larch. We estimate that most of the mature or larger larch in Alaska died in the beginning of the mid 1990s. They are dead now and we have relatively few larger or older larch surviving.

"That was the larch saw fly. Again it is the classic sort of insect. It's a moth that goes through these big outbreak numbers when there are warm conditions. This happened and persisted for a number of years. This assisted the insects in building up their population numbers. Then they just exploded across the landscape at huge population densities, found most of the mature larch and ate them up.

"In the first year of defoliation, heartier larch will survive. The next year they will grow out sometimes even more because the insects convert plant material that otherwise might accumulate in hard to decompose compounds, into nitrogen-rich droppings from the insects on the forest floor. But the second year of defoliation in a row, it weakens the health of trees and it is starting to kill the weak trees. Then, if they are defoliated a third year in a row, most of the trees die.

"That is exactly what happened. The outbreak was so severe, and so widespread, that most of the older larch died. And I could go on with other examples. The point is that associated with this unusually warm last couple of decades of the 20th century that we have had exactly what you would expect which is major outbreaks of the insects that affect trees and their outbreaks have been so severe and so persistent that they have caused unusually high levels of tree mortality to the point where we are talking about landscape changes.

"Spruce bark beetle probably have a limited potential to spread. The reason seems to be that when you go out to the coast, the spruce is a hybrid between sitka and white spruce. That is where most of the mortality is taking place. When you get into the sitka spruce stands, that is much more resistant to the insects. Sitka spruce can be killed by outbreaks of spruce bark beetle but much, much rarer, and usually they have to be severely injured by something that sets them up for it.

"As you go northward into the pure white spruce, it looks as if the cooler temperatures and the drier humidity limits the ability of the spruce bark beetles to build up in sufficient numbers so that there is probably a natural limit to the landscape level outbreaks to the north. It was just in this transition zone that they were uniquely vulnerable and the bugs just cleaned them out in that area.

"But there is another aspect to this. In addition to the already established insects that are usually under the control of colder temperatures, and now with the warmer temperatures have gone into these extreme outbreak levels, the warmer temperatures may be allowing insects that did not occur here before, and only south of here to become established. And we may have an example of that with the spruce bud worm. Rather than a beetle, it's a moth and instead of eating the bark, it tunnels into the new buds that are emerging.

"It is well known and well described through the southern boreal region of Canada. It is a major factor in boreal forests there. In general there is this temperature control that seems to help calm these big outbreaks, or limit it to background levels when it is cooler. The entomologists in Alaska didn't report it as an outbreak level in the boreal region here north of the Alaska Range before. And in the early 90s it first appeared in outbreak levels during a period of some of our warmest years. And it stayed around and appears now to be building up to new outbreak levels.

"In other words, it looks like it successfully established. This insect that can cause tree mortality has been established here. So that is another way in which the warmer temperatures are changing the forest. There are many places with tens of thousands of contiguous acres where every overstory tree is dead. It is a very impressive site."

Fairbanks Climatologist Uma Bhatt said, "I drove south of Anchorage sometime in July and you just see huge stands of dead trees. It is pretty dramatic. I've seen photographs from researchers from the air. It is pretty amazing. I was in Tennessee before I moved here and there was another beetle there that ate through the pine trees so there were acres and acres of pine that were dead. Those were all planted by humans (in Tennessee). In Tennessee there is a wide variety of trees that can grow. In Alaska there are very few trees that can grow. There isn't a lot of variety. So when one gets killed or when one gets sick it is very easy for it to spread to other trees of the same species."

Climatologist Bruce Molnia said, "Flying around Alaska, you see areas that were live, green needles that are now dead with brown needles. There is a major change."

Williams warned that "it is hard to imagine what (the dead forest) will be replaced by. They originally thought it might be replaced by aspen or birch, but when you see that they are being attacked by leaf minors, with leaves that are just chewed up due to leaf minors that are prospering because of global warming, you wonder what will succeed them. It is unbelievable to understand that in such a short period of time that we have had this devastation, and much of it was not anticipated.

"That is the other story about disease vectors. Again you can say 'yeah, glaciers melt, sea levels rise, whatever' but there are I think so many diseases that we can't even anticipate as global warming advances."

Williams has a cabin near Anchorage overlooking Matangska Valley and Glacier. "Most of our old spruce on our property has died from spruce bark beetle. We tried to make the best of it by building our cabin with dead spruce. But, to look out and see basically every mature spruce die in the course of about five years was gut wrenching. We also have quaking aspen. Last summer for the first time we found leaf minors in our quaking aspen. So our quaking aspen instead of having brilliant green leaves in the summer and beautiful yellow leaves in the fall, last summers' leaves were gray: gray. And in the fall they were brown as the leaf minor chewed up the leaves.

"Last year is the first year that we had it. Fairbanks had it a couple years before, but it is coming south in Alaska. And to think, all of this in our lifetime."

I spent an intensive month exploring the Tongass National Forest of Southeast Alaska by kayak, foot, Greenpeace boat (Arctic Sunrise), Greenpeace helicopter, ferry, car and bus, for an upcoming book. I realized deeply the need to protect the old growth forest which comprises the largest tracts left in America, supporting an amazing ecosystem with tens of millions of salmon, 30,000 brown bear, gray and humpback whale, sea otter, seal, tens of thousands of bald eagles and a vast array of other species, along with the Tlingit, Haida and Tsimshian native peoples.

The ancient sitka spruce, western hemlock and western red cedar tower 200 feet high in the remaining uncut stands. About 70 percent of the ancient forest still remains in the Tongass, and the ecosystem still has its integrity. But logging still threatens this marvel of nature until all of the old growth forest is permanently protected.

In Juneau, I joined a feast with many of the leaders in the effort to protect the Tongass, sponsored by the Alaska Coalition. Over salmon and wine, we looked out upon the Mendenhall glacier while more fish roasted on an open fire. The glacier was retreating fast, leaving a barren glacial moraine devoid of life. The trees grew taller farther away from the glacier, reading of the past glacial recession. Using such records, as well as measurements of glacial mass over the last several decades, scientists can piece together a picture of how fast glaciers are melting. Their findings are startling.

With the increase in temperature, glaciers in Alaska are receding fast. A team of Alaskan researchers found that the rate of thinning has doubled in the past five years compared to the prior 40 years.

"Glaciers are formed by overflow of snow in the wintertime," said Gwich'in elder Sarah James. "Some of the glaciers get thawed out by the end of summer. Snow would be melting, there would normally still be snow on a glacier at the end of summer, but now it all melts."

The Columbia Glacier retreated nearly 13 kilometers between 1982 and 2000. In 1999 the retreat rate increased from 25 meters per day to 35 meters per day. The glacier now retreats at a rate of half a mile per year (.8 kilometers).

"The Columbia Glacier in southern Alaska began a catastrophic retreat in the mid 1980s," reported Michael Hambrey and Jurg Alean in "Glaciers," (1992). The retreat has caused a massive increase in icebergs. The prediction that these icebergs would be a hazard for boating proved true when the Exxon Valdez, carrying oil from the north slope of Alaska that had been pumped to Valdez through the Trans Alaskan Pipeline, tried to avoid ice from the Columbia glacier. While it avoided the ice, it ran aground on Bligh Reef (with the contributing factor that thc captain was drunk), spilling 11 million gallons across more than a hundred miles of some of the most environmentally sensitive areas of the state. The contamination is still causing massive ecological impacts more than 15 years later.

The melting of North America's largest and longest glacier, the Bering Glacier, was found to be twice as fast as previously thought (Detroit Free Press, May 12, 2007). Representing 15 percent of the ice in Alaska, that glacier now releases 8 trillion gallons of water a year into the ocean – twice as much as the flow of the Colorado River.

"Ninety-nine percent of the temperate glaciers under a mile above sea level are receding, thinning or stagnant," said Glaciologist Bruce Molnia. "Most of those that are stagnant are insulated by rubble on top of them. Temperate glaciers are those that have both ice and water. They make up three percent of glacier ice and are the most sensitive to global warming." He said glaciers north of the Arctic Circle are melting as well. "The McCall Glacier in the Brooks Mountain Range has been melting about a foot a year," Molnia said. "Most of the glaciers in Alaska are wasting away." Glaciers in Glacier Bay have retreated 60 miles since the 1800s.

"Polar ice, which is far below freezing, is a lot less sensitive. It may change from -60 to -50 degrees but it still will not melt," Molnia said. "Polar ice is accumulating in Antarctica—it recently covered the U.S. base on the South Pole. But the edges of Antarctica and Greenland are melting."

Guntner Weller said, "The thousands of glaciers in Alaska have seen dramatic losses. Some of the smaller glaciers are disappearing entirely, as is happening throughout the world, like Glacier National Park."

Glacier National Park had more than a hundred glaciers in 1900. By 1964, it was down to 67 glaciers. In 2001, there were 37 remaining. In early 2008, 27 remain. Scientists predict there will be none left by 2030. I routinely backpack and fish at the park, because Amtrak's northern cross-country train, the Empire Builder, makes two stops at the park.

When not fending off giant grizzlies from my fresh-caught fish, I have walked the vast stretch of moraine where just a few decades ago there had been a vast glacier, next to remnant ice packs that were recently named glaciers in the park. The change is vivid. Glacier Park rafting guide Denny Gignoux said, "you see the changes quite rapidly."

In Alaska, it is Glacier National Park on a far grander scale. "There are thousands of glaciers in Alaska and they are receding," Weller said, adding that melting from the Alaskan Glaciers adds the amount of water that would cover an area the size of Alaska five meters deep to the ocean annually. "It is a significant contribution to sea levels. The reduction in thickness and length in recent years have accelerated. Scientists fly over the glaciers and use a laser altimeter to measure it. As they do that from year to year, they can measure the water loss."

Between 1850 and 1960, glaciers retreated 7.5 percent in Alaska. Between 1960 and 2000, there was a 7 percent retreat, reported Ohio State University Geological Science Professor Dr. Lonnie Thompson in the New York Times, November 9, 2004. This is a rate of increase of nearly three times. The professor also documented large puddles of melting ice at 20,000 feet in the Himalayas, where it had been frozen for thousands of years. The melting is creating new lakes in the Himalayas as water accumulates behind dams of ice that could break, unleashing flash floods.

You can see the scale of melting in walking up to the massive glaciers. Their immensity, and the hundreds of yards of barren ground that had just been glacier the year before, are awakening people who frequent glaciated regions to the severity of global warming.

"I have lived up here for 23 years," Williams said. "Just going to Prince William Sound and seeing tidewater glaciers that you have experienced a mere 10-15-20 years ago, you come around the bay and their retreat, their diminishment is breathtaking. It is like someone punched you in the stomach when you see these glaciers that once commanded this extraordinary presence in a bay or inlet but now are diminished and small. In a fraction of a lifetime to see that change again and again is incredible.

On Kilimanjaro in Kenya, an 11,700 year old ice cap that was 4.3 square miles in 1912 had retreated to .94 square miles in 2000, losing 92 percent of its former ice by 2007. Scientists predict that the ice cap will disappear by 2015. "When you see the big picture accumulating from many sites, the evidence of drastic climate change becomes quite compelling," Dr. Thompson said.

In addition to raising sea levels, glaciers can also increase the frequency of earthquakes in this, one of the most seismically active places on Earth. "Retreating glaciers in Southern Alaska may make earthquakes more likely there, according to a study by NASA and US Geological Survey Scientists." The paper reported that glaciers in Alaska had shrunk at least ten percent in the last hundred years. "As they melt, they lighten the load on the Earth's crust and allow the slowly drifting tectonic plates that make up the crust to move more freely. The collision of two plates under the Pacific Ocean off Alaska creates a buildup of pressure, which is relieved by earthquakes."

"This just makes it easier for earthquakes to occur," the Post quoted NASA Geophysicist Jeanne Sauber. "This is equivalent to years of tectonic strain accumulating." Sauber and Bruce Molnia co-authored a study revealing these findings in the July 2004 issue of the journal Global Planetary Change.

This is no laughing matter in what scientists say is the most earthquake-prone place on Earth. An Alaskan state brochure on earthquakes reads, "It is no surprise to anyone who has lived in Alaska a few years that our state is one of the most seismically active regions of the world. Most of us have felt earthquakes at one time or another." Three of the world's six largest earthquakes have occurred in Alaska since 1900. Earthquakes of a magnitude of 7 or greater have shaken some part of the state at least once a year; quakes of a magnitude of 8 or greater have taken place once every 13 years on average.

In 1964, an earthquake of a magnitude 9.2 hit near Anchorage, killing a hundred people in Alaska and 16 people on the coasts of California and Oregon, mostly from the massive wave, called a tsunami. The giant wave triggered landslides and large scale destruction.

The phenomenon is not constrained to Alaska, but is worldwide. Greenland contains the largest mass of land-based ice outside of Antarctic, with eight percent of the world's ice. There, the ice sheet has thinned more than a meter per year on average since 1993 along parts of its southern and eastern edges. University of Colorado at Boulder scientist Konrad Steffen found that 16 percent more Greenland ice melted in 2002 than in 1979, when satellite monitoring began. The rate doubled from 2002 to 2007.

The same warming weather around the far north is being felt in the far south as well. Scientists blame global warming for the disappearance of child-sized Adelie penguins from the northern tip of the Antarctic Peninsula, which juts out toward South America. The pack ice has become so scarce that the penguins can no longer use it as a resting platform for their winter fishing.

Antarctica, with 91 percent of Earth's ice and an average of 2.3 km. in thickness, is also melting. In the 1990s, three ice shelves along the edges of the Antarctic Peninsula fully disintegrated. Two more ice shelves lost more than one seventh of their combined 21,000 square kilometers between late 1998 and March 2000, an area the size of Rhode Island.

Worldwatch Institute reports that as a whole, the world's glaciers are now shrinking faster than they are growing, and losses in 1997-1998 were "extreme," according to the World Glacier Monitoring Service.

Scientists predict that up to a quarter of global mountain glacier mass could disappear by 2050, and up to one half could be gone by 2100. In other parts of the world, Himalayan glaciers are expected to shrink by one-fifth, to 100,000 square kilometers in the next 35 years. In Nepal, melting ice prompted a glacier lake to burst in 1985, sending a 15-meter wall of water rushing 90 kilometers down the mountain, drowning dozens of people, and destroying 14 bridges and hundreds of homes. Scientists warn that now such disasters are much more likely.

Ice-melt on land increases sea levels as it flows into them. Over the past century, melting in ice caps and mountain glaciers has contributed on average about one-fifth of the estimated 4-10 inch rise in global sea levels. The rest is caused by the thermal expansion of the oceans as the Earth has warmed. But the share of ice melt in the rise of sea levels is increasing, and will continue to accelerate if the larger ice sheets crumble as expected far sooner than scientists thought just a few years ago.

The increase in water to sea levels threatens to flood low lying areas such as New Orleans (again), Florida, New York, Tokyo, Shanghai, Bangladesh, islands throughout the planet, much of the Netherlands and more.

Chapter 4

Ice Sheets Accelerating Flow Into Sea

Far greater in size than glaciers, ice sheets hold by far the vast majority of Earth's ice. Formerly thought to just melt slowly, affixed somehow to the ground, the vast ice sheets of Greenland and Antarctica are accelerating their flow to the sea, as melting rivers are plummeting through crevasses to the bottom of the 2-3 mile-thick ice sheets, then flowing under the ice and lubricating their slide into the Ocean. Once in the sea, the ice sheets break up, and giant icebergs float with ocean currents to more temperate regions where they interfere with shipping and oil and gas development, melt, and raise sea levels.

The largest ice shelf in Canada, the Ward Hunt shelf, broke up between 2000 and 2003 with the 1.8 degree Celsius average temperature in the region being to blame. This left six large ice shelves in the area. On August 13, 2005 that became 5. The 3,000-year old Ayles Ice shelf collapsed, causing tremors measured 250 kilometers away. It has now floated toward the North Pole, and is slowly melting. It is expected to drift into the Beaufort Sea in coming years, where it may pose a hazard to shipping and oil and gas development. Though the Canadian government detected the tremors from the ice shelf breakup at the time, they did not announce it until December, 2006, after they had time to study it.

Arctic Screaming

While scientists once thought Greenland's ice sheet might take a thousand years to melt, longtime NASA Goddard Institute for Space Studies Director James Hansen reported that these studies are junk in the face of new findings. Glaciers are moving three times faster into the sea than they were just ten years ago. Hansen announced in February, 2006 that the rate of melting in Greenland had doubled in the last 5 years. From underneath the melting ice sheets, Warming Island emerged in 2007 off of Greenland.

Following the current path of the world's course, with increasing emissions of greenhouse gases, Hansen said we would likely see a three-degree Celsius temperature increase this century at a minimum. The Intergovernmental Panel on Climate Change (IPCC) reported in 2007 that global average temperatures are expected to rise up to 6.4 degrees Celsius by 2100.

But the report was released just before the Global Carbon Project 2007 analysis that found global carbon emissions are rising much faster than the rate used by the IPCC analysis. From 2000 to 2004, emissions rose three percent a year: triple the rate during the 1990s (Christian Science Monitor, May 22, 2007). In a separate report, the U.S. projects that global carbon dioxide emissions will increase by 59 percent from 2004 to 2030 (Reuters, May 21, 2007).

A study by the Australian Greenhouse Office released May, 2006 found that the warming could move very rapidly, trending toward or beyond the higher end of the range established by the IPCC. They found that carbon sinks (forests, permafrost, and other lands with vegetation), are likely to turn into carbon sources. 2007 reports found the boreal forests are beginning to release carbon faster than they absorb it. Two studies published in Geophysical Research Letters, May, 2006, also found that temperatures at the end of the century could be significantly hotter than current climate models predict. They predicted temperatures could reach eight degrees Celsius warmer than 2006.

The last time that the Earth was three degrees Celsius warmer was during the Pliocene era, when sea levels averaged 25 meters higher than today -- 80 feet -- Hansen reported in February, 2006. We would live in a different world, the scientist said, unless we make major changes within a decade. The rate of warming today is faster than it has been for millions of years.

Greenland is now losing about 200 cubic kilometers of ice per year. "Hundreds of cubic kilometers sounds like a lot of ice," Hansen wrote. "But this is just the beginning. Once a sheet starts to disintegrate, it can reach a tipping point beyond which break-up is explosively rapid. The issue is how close we are getting to that tipping point. The summer of 2005 broke all records for melting in Greenland. So we may be on the edge." Hansen wrote this in February, 2006. Since then, the summer of 2006 exceeded the melt of 2005, and 2007 was off the charts.

"Seismic stations reveal a shocking increase in icequakes on Greenland, caused by a portion of an ice sheet lurching forward and grinding to a halt," Hansen reported. The number of icequakes registering 4.7 or greater on the Ricter scale doubled from 7 in 1993 to 14 in the late 1990s. It doubled again by 2005.

Like the Pentagon predicted, climate change may have been seemingly gradual thus far, but it appears ready to burst into an onslaught of Katrina-like tragedies that will occur with ever increasing frequency.

The melting of Greenland's ice sheet would raise sea levels by 23 feet. And then there is Antarctica, with 11 times as much ice as Greenland. The Larsen B Ice Shelf, which was stable for 12,000 years, suddenly collapsed in 2002, with 3,250 square kilometers of the shelf breaking off within 35 days. The ice disintegrated, floated to warmer waters and melted. Scientists thought it would take centuries for the ice to melt. Over the last 5 years, a total of 5,700 square kilometers have broken off of the Larsen B ice shelf.

The Wilkins Ice Shelf on the southwest Antarctic Peninsula has started to collapse, reported the National Snow and Ice Data Center March 25, 2008. Satellite imagery shows that the 13,680 square kilometer (5,292 square mile) ice shelf has begun to collapse because of rapid climate change in the fast-warming region of Antarctica. The agency found "runaway disintegration of 405 square kilometers (160 square miles)." Western Antarctic has warmed almost 5 degrees Fahrenheit in the last 50 years.

Melting particularly accelerated in 2007 in the Antarctic, reported the head of Argentina's Antarctic Institute's Glaciology

Division. Thinning of glaciers in West Antarctica is 20 times faster than previously (Reuters, March 14, 2008).

Ohio State University Professor Lonnie Thompson also reported that melting glacial water is pouring through the large crevasses in the ice sheets. The water then flows underneath the ice sheets, allowing the entire ice sheet to flow into the sea, at an ever increasing pace. "Once these ice streams start moving, their influence stretches right to the interior of the ice sheet," Hansen wrote. "Building an ice sheet takes a long time, because it is limited by snowfall. But destroying it can be explosively rapid."

"The business-as-usual scenario, with five degrees Fahrenheit global warming and ten degrees Fahrenheit at the ice sheets, certainly would cause the disintegration of ice sheets," Hansen said. "The only question is when the collapse of these sheets would begin. The business-as-usual scenario, which could lead to an eventual sea level rise of eighty feet, with twenty feet or more per century, could produce global chaos."

Antarctica has 70 percent of the planet's fresh water. The collapse of the 360,000 square mile West Antarctic Ice Sheet (WAIS), an ice mass the size of Mexico, would raise sea levels by an estimated 6 meters (Nature, 1998). The melting of both of Antarctica's ice sheets would raise global sea levels by 70 meters. This is different than the melting of Arctic Ocean sea ice or the floating Antarctic ice shelves, which have no effect on sea levels because they already displace water.

Arctic Screaming

With icebergs from Antarctica having been reported off New Zealand, and the massive avalanche of new data on climate change, no credible scientist now questions that climate change is raising sea levels, with more to come. The debate is over how fast the increase will be, but no matter what, it will be devastating with every inch of rise, with billions of people living in coastal areas.

NASA Scientist Jim Hansen said that to keep the increased warming under two degrees Fahrenheit will require a major change in direction this decade. Looking historically at when the earth was two degrees Fahrenheit warmer than today was 130,000 years ago, during an interglacial period. Then, sea levels were six meters higher: nearly 20 feet.

Al Gore showed what our country, and some of the world would look like with a 20-foot sea level rise in his movie "An Inconvenient Truth." It is not a pretty picture. Eighty feet of sea level rise, as Hansen warns could happen with business as usual, is downright alarming. While it does not appear that this will occur this century, by raising the temperature, and continuing on the path of ever increasing greenhouse gases, sea levels most likely will rise to that point, whether this century or how many more being unclear. What a terrible legacy to leave future generations.

In a review of Al Gore's "Inconvenient Truth," Roger Ebert wrote, "In 39 years, I have never written these words in a movie review, but here you are: You owe it to yourself to see this film. If you do not, and you have grandchildren, you should explain to them why you decided not to (Chicago Sun Times, June 2, 2006)."

The dynamic nature of the ice sheets, and the alarming pace of increased warming, far faster than scientists had predicted, makes it clear that this is the issue that we must address now. The threat to our survival is greater than any other facing the planet today, short of a full scale nuclear war.

Chapter 5

North Atlantic Current Slowing

Not only is this melting ice raising sea levels, it is also disrupting critical oceanic currents that help to moderate our planet's climate. Earth's oceans play a major role in regulating the global temperatures of the planet. Each ocean has deep currents that moves cooler polar water towards the equator, pulling back warmer water that helps to moderate global temperatures.

Scientists are finding that the thousand-mile wide North Atlantic Ocean current is very complex and very sensitive to temperature and other climatic changes. The melt-off of fresh water in the Arctic may shut down the North Atlantic Oscillation that brings warmer tropical water to Europe and the northeastern United States, keeping England, which is in the same latitude as Labrador, about ten degrees F. warmer.

Raymond Schmitt, a Senior Scientist in the Department of Physical Oceanography at the Woods Hole Oceanographic Institution said in 2000, "the crucial role of the oceans in climate has not been sufficiently acknowledged in most research on climate change to date, including the National Climate Assessment Report under discussion here. It was a tradition of the climate modeling community to treat the ocean as a shallow swamp; a source of moisture, but playing no role in heat transport and storage. We now know this to be a significant error. The oceans are an equal partner with the atmosphere in transporting heat from the equator to the poles, and a reservoir of heat and water that overwhelmingly dwarfs the capacity of the atmosphere."

Schmitt emphasized the importance of Oceans to global weather patterns, noting that they:

- "Cover 70% of the surface of the Earth.
- "Have 1,100 times the heat capacity of the atmosphere (99.9% of the heat capacity of the Earth's fluids)
- "Contains 90,000 times as much water as the atmosphere (97% of the free water on the planet)
- "Receive 78% of global precipitation"

As salty water in the North Atlantic gives off its heat, from having traveled from tropical areas, it sinks. This pulls warmer surface water from the south, creating the North Atlantic Oscillation that has a massive impact on temperatures around the Atlantic. Schmitt said that this effect, called "thermohaline circulation," is far greater than the amount of water moved through evaporation and precipitation, and is the largest source of ocean transportation of heat from the tropics to the Arctic. This effect, also called an ocean conveyer, helps to pull the warmer Gulf Stream waters into the North Atlantic. The slow current moves more than a hundred times the amount of water as the Amazon River, warming Europe by about 18 degrees Fahrenheit. This effect helps to moderate winter conditions throughout the North Atlantic region.

Arctic Screaming

Scientists measure the ocean currents with drifting monitors like weather balloons that flow with the current then rise to the surface to report salinity and temperature data to satellites. Then they drift another 10 days, repeating this up to 150 times. They have found that it is slowing down. Scientists report that the ocean circulation has slowed by about 20 percent, the Seattle Post Intelligencer reported April 16, 2004.

If the waters of the North Atlantic become too fresh from the melting of glaciers and increased precipitation, the water may not sink, shutting down the oscillation. Schmitt reports that this has happened before. The most recent one occurred about 12,000 years ago, he said, with very dramatic consequences on the northern hemisphere, returning much if it to glacial conditions for a thousand years. Fossil records indicate that this happened in only 10-20 years.

The temperatures of the North Atlantic were five degrees Celsius (nine degrees Fahrenheit) cooler than previously. Because of the greenhouse effect that is increasing with our emissions, along with the other positive feedback loops, such massive cooling is unlikely today, but this could cause Europe to warm at a slower rate than the global average.

Scientists released a study in December, 2003 that found a significant increase in salinity in the tropics, and a decrease in salinity in areas of higher latitudes, indicating it may not be long before the North Atlantic Oscillation shuts down.

In addition to the overwhelming weight of evidence for this phenomenon, NASA reported in April, 2004 that they had found a slow-down of the North Atlantic Oscillation in the late 1990s from analyses of satellite data. Co-author of the study Peter Rhines warned that what happens in the North Atlantic affects global weather patterns. "It's like Grand Central Station there, as many of the major ocean water masses pass through from the Arctic and from warmer latitudes," he said. The North Atlantic Oscillation, also called a subpolar gyre, can take 20 years to complete its cycle. Warm water runs north through the Gulf Stream, past Ireland, before it turns west near Iceland and the tip of Greenland.

Scientists estimate that the overall warming will eclipse any cooling in the North Atlantic, but only time will tell. Overall, the earth will certainly warm. But where it warms the most, and how much different regions warm, is more complicated.

A paper coauthored by Terrence Joyce, Senior Scientist, Physical Oceanography and Lloyd Keigwin, Senior Scientist, Geology & Geophysics found that "Evidence for abrupt climate change is readily apparent in ice cores taken from Greenland and Antarctica." Joyce and Keigwin wrote that the North Atlantic Oscillation "heat engine operates to reduce equator-to-pole temperature differences and is a prime moderating mechanism for climate on Earth." Fresh water from glacial melt-off can stop or reduce the effect of this mechanism.

The two scientists wrote, "Recent evidence shows that the high latitude oceans have freshened while the subtropics and tropics have become saltier. These possible changes in the hydrological cycle have not been limited to the North Atlantic." Another impact of this oscillation is by increasing precipitation over the poles, where it stores as glacial ice, affecting the hydrology of the planet. Shutting it off might lower this precipitation, thus decreasing the amount of snow over the North Pole.

Shutting down the current could also turn the tropical rainforests of South America into desert landscapes. The more we learn, the scarier climate change becomes. We are messing with a volatile system far more sensitive to our impacts than we realized just a decade ago.

Chapter 6

Climate Change Attacks Gray Whales, Salmon

From Juneau, a friend of mine from Chicago, Joe Jensen and I took a ferry to Sitka, where we launched a double-Klepper kayak. The kayak folds up, and a friend of his in Oregon had it mailed to Alaska. We quickly figured out that there was only enough room for one of us, and Joe graciously gave the adventure to me so that I could bring the findings to the masses through this and an upcoming book devoted just to the need to protect the Tongass.

Kayaking in the region, I felt the bounty of life with the tug on my fishing rod, the sea otter swimming acrobatic moves below in clear water and the whales surfacing nearby. But even in this magical place so teaming with life, climate change is hammering the ecosystem, as is the chainsaw.

The whales that I admired as they surfaced dozens of feet away (luckily not right underneath my kayak), recently had a plunge in their population. Gray whales experienced a major dieback in 1999 and 2000 that scientists are just now realizing was likely caused by global warming in the Arctic. Their population went from about 26,600 to 17,400. Gray whales feed on shrimp-like crustaceans. The crustaceans feed off algae that live on the bottom of the ice. The ice retreated far earlier than previously, and so did the crustaceans that the gray whales swam up to the Arctic to feed upon.

Because of the lack of food, whales have begun to stop to feed on their way north, trying new foods which can cause parasites and other hazards to the whales. Near San Francisco's Golden Gate bridge, whales feeding on the sea floor were ripped apart when they came up to breathe, and were hit by the propellers of cargo vehicles. Other whales have been extending their 5,000 mile northward migration through the Bering Strait and Chukchi Sea into the Beaufort Sea.

This has surprised Inupiat hunters, and is now interfering with their subsistence. Inupiat are allowed to hunt bowhead whale, but not gray whale. As they see the spouts from gray whales, they rush over thinking it is a bowhead, but only find gray whales, which they are not allowed to take. They have also hit gray whales with their skiffs.

The whales feeding before me as I kayaked through the islands of the Tongass Forest that early July would next be heading around Alaska's coast and all the way up the Bering Strait. Many would swim into the Arctic Ocean, where I too would venture on my long journey ahead, though along a much different route.

As it was growing near dark, I came upon several grizzly bears munching on salmon by a small stream that entered the channel. In just a few feet of water, salmon were jumping about in all directions. I cast past where one had just leapt from the water, using a three-inch Rapala. Soon, I had a feisty Pink Salmon on my line. The fish pulled me toward the spawning creek – and the bears. Though I tried to get a few paddle strokes in, while holding the rod at the same time, the salmon prevailed in pulling me closer and closer to the bears.

Soon, I was about 15 feet away from three large grizzly bear munching on salmon. One of them glared up at me. Fortunately, the salmon then swam around to the other side of the kayak, and began pulling me away.

Hundreds of bald eagles perched in giant trees on the mountainside. Eagles swooped down near my kayak and plucked up salmon, then circled around and around, slowly lifting the fish with great effort, before taking them back to a tree perch to eat.

Even in this remote area, boats regularly traveled through the inside passage, along with the giant ferries and cruise boats. Cruise boats left a trail of brown pollution for miles behind them, and routinely dumped their trash and raw sewage right into the water. That, along with container ships plying the waters, and fishing vessels showed that we are burning large amounts of fossil fuels in even the remotest areas of our planet. About a million people a year come to this area, mostly on cruise boats.

Climate change is devastating southern Alaska, as well as its horrific impacts in the Arctic region where I was ultimately headed. Just before my kayak adventure, Joe Jensen and I had a halibut lunch with Greenpeace volunteer Larry Edwards in Ketchikan. He reported salmon kills in the area because of increased temperatures. "The summers are getting hotter because of climate change,"

Edwards said. "So here in the South Tongass, we are having problems with the overheating of salmon streams. Last summer, there were fish kills at Stanley Creek on Prince of Wales Island."

Another climate change caused impact is increased predation by salmon sharks. Alaska Conservation Solutions President Deborah Williams said salmon sharks are on the increase, threatening salmon populations in southern Alaska. "I was on the Trustee council for the Exxon Valdez oil spill. And we were going around an island in The Gulf of Alaska, because we were looking at using some of the settlement money between Exxon Valdez and the state of Alaska to buy some land on Afognak Island. We went to this one bay and it was filled with somewhere between a hundred and two hundred salmon sharks.

"Salmon sharks are thriving in the warmer water. I've only seen a handful of salmon sharks in my 23 years in Alaska. And to just go in this bay and see potentially over a hundred salmon shark fins in this bay, it was breathtaking in a very negative way." Salmon sharks threaten the thriving salmon fisheries in Alaska that have evolved largely without their predation.

Globally, salmon have been hammered by climate change and other impacts such as disease from "farm" raised salmon grown in massive containing nets in the ocean. The Environment Agency of Wales reported that rod catches of salmon declined more than 95 percent, from 7,684 in 1967 to 357 in 2002 on the River Wye. Scientist Rob Evans with the agency's fisheries division said that "the main reason for the decline (in salmon) across Central and Northern Wales has been climate change in the northeastern Atlantic."

Scientists attribute climate change to the fast decline of the Kenai River red salmon in southern Alaska. The salmon spawn in Skilak Lake, which is fed by glacial melt. With the rapid increase in this melting, silt increased the turbidity of the water by 65 percent, and zooplankton declined 50 percent, reducing the food supply of the young red salmon. The salmon fry in 2006 were 60 percent below normal size because of the reduced food, and smaller fry have a harder time surviving.

Just with a lure out the back of my kayak while I paddled, I hooked salmon after salmon. Often within a minute I would have a fish. Because I generally do not just throw the fish back, but only catch that which I intend to eat, I did not do much fishing. But I experienced their abundance anyway as they jumped out of the water all around, and were hoisted into the air by bald eagles. The amazing abundance is awe-inspiring, but very fragile, as has been seen nearly everywhere else on the planet, that once had such abundance.

But even here, our reach goes far beyond the fish being hauled in by the many fishing vessels that I watched plying the waters of the inside passage. By increasing temperatures, wildlife, both aquatic and terrestrial, are moving north in the Northern hemisphere, and south in the Southern hemisphere, to try to keep up with the changing climate. The massive change is alarmingly disruptive, and destructive of the abundance that has evolved to fit the former stable interglacial time period.

"Fisheries are a big part of Alaska," said Climatologist Guntner Weller. "We export a lot. The effects of climate change on fisheries could be quite dramatic, too. It is clear that ocean temperatures have an impact on the distribution of fish. They are very sensitive. Fish are displaced (as waters warm) and new species come in. This could mean a relocation of infrastructure. The biggest fishing ports are Kodiak and Dutch Harbor (on the Alaska Peninsula). The fishing ports may need to change. It is a huge impact on fish supplies for the U.S."

Chapter 7

Climate Change Threatens Oceans

In addition to changing the location of fish, which would likely cause a massive collapse, increased carbon dioxide presents an even more ominous threat: acidification. A team of 50 scientists from 9 countries found that the increase in carbon dioxide in the atmosphere has made our oceans measurably more acidic since the industrial revolution. The scientists warned that this is a serious concern in a report released in July, 2006.

"Unlike any possible controversy over global warming, as you increase carbon dioxide in the atmosphere, you're driving carbon into the oceans and increasing the ocean's acidity – and this is not debatable," Joan Kleypas, an ecologist and geologist in Boulder Colorado told the San Francisco Chronicle.

The government-sponsored scientists found that this increased acidity threatened the widespread destruction of shell-building organisms that form the base of the marine food chain. These organisms are responsible for building coral reefs. The scientists reported that billions of tons of carbon have poured into the oceans from the burning of fossil fuels, causing "the most dramatic changes in marine chemistry in the past 650,000 years," said federal oceanographer Richard Feely in a San Francisco Chronicle article July 6, 2006.

Ice cores show that the ocean acidity level has remained steady for hundreds of thousands of years. But the accelerated burning of fossil fuels dumped 118 billion metric tons of carbon from the atmosphere into the oceans between 1800 and 1994. And the pace of that change is increasing along with the amount of carbon dioxide in the atmosphere. A total of 142 billion tons were dumped into the oceans as of 2006, with more than 2 billion tons entering the oceans each year.

Shell building organisms, known as marine calcifiers, include some of the most abundant and important food sources in the oceans, including plankton that form the base of the food web, feeding salmon, mackerel, cod and countless other animals.

National Oceanic and Atmospheric Administration Engineer Christopher Sabine said the findings were "a call to arms." Our continued path of increasing greenhouse gas emissions would increase the acidity of the oceans by 150% by the end of the century, with potentially disastrous consequences.

Increased acidity reduces the ability of marine animals to calcify and build shells or calcium carbonate skeletal structures. Dr. Victoria Fabry said "The potential is there to have devastating impacts (Anchorage Daily News July 6, 2006)." Already, calcification rates of a common coral species in the Great Barrier Reef have slowed 21 percent in the past 16 years (London Times, February 23, 2008).

Like the massive disease outbreaks that climate change is causing, the most frightening impacts of climate change are just now being discovered. Because we are undergoing a giant experiment, increasing greenhouse gases in the atmosphere at least 30 times faster than ever in the Earth's past, we do not know what the exact impacts will be. But we do know that the negative impacts vastly outweigh any seemingly positive benefits of global warming.

Fish globally are dying because they cannot handle the fast rise in temperature as well. Dead fish are washing up in Qatar, where water temperatures in August were 36 degrees Celsius when they should have been at most 32. The Supreme Council for Environment and Natural Resources found no evidence of bacteria or oils, indicating it is the higher temperatures that are to blame.

Ted Stevens issued a press release in 2007 requesting targeted research on the impacts of climate change and ocean acidification on fisheries. This, along with John McCain and Arnold Schwarzenegger's support for climate change action shows that this is not a partisan issue.

From Sitka, I took the ferry to Hoonah. There, I could see the whole mountainsides stripped bare of trees. Right from the small town, nearly every steep mountainside on both sides of the channel were stripped bare of trees. As I was interviewing some local residents about the eyesores that once were old growth forests around their town, a young man asked, "what do environmentalists use for toilet paper?" I said that I used 100 percent post consumer waste recycled toilet paper, made from recycled newspaper. The kid said that he had just seen that on a bumper sticker.

The irony is that much of the amazing virgin forest was logged to make toilet paper. To trade such an amazing ecosystem, that supports the vast abundance of salmon, bears, bald eagle, whale and native peoples for toilet paper that could have come from recycled paper, of which there is no shortage, is insane. The old growth forest supports more than 30 times more jobs in tourism and fishing than the few logging jobs. Many times more logging jobs would be created in select-cutting second growth areas, and saving all remaining old growth. Much better is to focus on re-using wood now thrown into landfills, and building furniture and products that last, rather than throw-away junk that has become the mainstay of "modern" society.

Recycling is another way to reduce greenhouse gas emissions, as well as resource consumption that cause the decimated clear-cut mountainsides so clearly visible when you get off the cruise line routes. Taxes should be shifted from income and property, which are good things, to toxic waste production, as well as environmentally destructive products, such as non-recycled paper, which we want to reduce. That way those who cause the problem pay the bill for government, which should increasingly work for solutions.

Needed is legislation mandating near 100 percent post-consumer waste recycled content in all materials possible. Doing that would all but eliminate the destruction of old growth forests that now are often clear-cut to make toilet and tissue paper. From Hoonah, I boarded the ferry for Haines, then Skagway, Alaska.

In Skagway, I watched salmon traveling up tiny streams I could step across. What an amazing abundance of life that once existed in most places on earth. We forget that salmon and other migrating fish once filled rivers all throughout our planet, offering sustainable feasts for predators and people on even the smallest creeks. But most have been stopped by dams or fished to extinction, or near extinction. As long as we can keep climate change from getting too out of hand, we still have an opportunity to restore many of the great fisheries.

From Skagway, I took the bus to Whitehorse, Canada through mountains with white-water creeks. Stakes 15-feet high with red tops rose from each side of the roadway for snow plows to see the location of the road. This was the route that thousands of Americans climbed over to reach interior Alaska in the late 1800s to mine for gold, in which hundreds died.

The Yukon River flows a clear blue in its headwaters by Whitehorse, but even thousands of miles upstream it is a large and very fast river. Seeing the current, I yearned to be in a boat, floating down the swift current and white water. The next summer I did just that in a canoe.

Rosen reported in the Christian Science Monitor October 7, 2003 that temperatures in rivers in Alaska are 5 degrees warmer in the last 20 years, making them nearly lethal to salmon. "Warming may have dire consequences for salmon in the Yukon River, the major food source for indigenous people along the 2,300-mile waterway."

Whitehorse is a small quiet community, with a scenic downtown and many bars. I walked to the salvation army thrift store to get some $6 used boots that did not properly fit, as mine had broken open with many holes.

There, I met Harvard Ayers and the group I would help guide through the Arctic Refuge. We had dinner together in Whitehorse. The next morning, we flew to Dawson, then to the Gwich'in community of Old Crow in the Yukon Territories, on the Porcupine River.

A smoky haze drifted across the forest, lakes and wetland bogs that extended as far as the eye could see from our plane. Millions of acres of forest were on fire in Alaska, and climate change was the primary cause.

Chapter 8

Gwich'in wisdom warns of major changes

When it comes to credible sources on a changing climate, Native Americans and scientists are a good choice. Indigenous peoples still living in an intact culture have knowledge passed down for thousands of years. They know that a lake freezes so deep by what date because they drill through it to fish. They know that the lake has frozen to what depth for hundreds or even thousands of years because their ancestors have fished that same lake, and passed down that knowledge. Scientists methodically work to document changes, with the scientific method ensuring strict adherence to the facts. When both groups sound the alarm, we should take notice.

Living off the Porcupine Caribou Herd, the Gwich'in People of the far north have maintained their culture while adding modern amenities to make life easier. Their deep knowledge of the land, friendly ways and loyalty to their people and the caribou that sustain them is awe-inspiring. At Old Crow, we were met by Chief Joe Linklater. The previous September, Chief Joe and I had met with Michael Dewine, who was an Ohio Senator at the time, thanking him for his efforts to protect the refuge.

Our group was honored to attend the annual General Assembly for Old Crow, held at Tlo Kut. The General Assembly gives residents a chance to directly question tribal leadership. With caribou, moose, salmon, duck, geese, and ample food always available, along with coffee and tea, all focus for days was the issues at hand. Climate change was a major topic of the General Assembly, and there was no question that it was harming the Gwich'in way of life and the greater environment upon which they depend.

At the meeting, there was much talk of climate change, and I did not have to ask about it. The Porcupine River had broken up much earlier than usual, in May, and taken along with it several Gwich'in meat caches and a cabin, luckily with no one in it. Everyone pointed to climate change as the cause.

The village government is putting in a permafrost research station to monitor the changes. Everything in the circumpolar Arctic is built on or depends upon the stability of the permafrost -- ground that is frozen year-round. With climate change, the permafrost is melting, causing everything from trees to buildings to collapse.

Cindy Dixon, who works at the Council of First Nations, spoke at the General Assembly. "Climate change is becoming a larger issue around the circumpolar Arctic," she said. "We're finding not just through science, but through our elders, our climate is changing. The migratory patterns are changing, the ice caps are melting. Inuit villages will have to move. Species are moving in that have not been there.

"The elders have for years been talking about the changes, and now scientists are documenting it." Elders reported that Musk-ox were found up in the mountains, which they said was caused by climate change. "We started to see a lot of animals and plants they've never seen," said Robert Bruce, Chief of Old Crow from 1991-1996. He said they drill the ice in the winter and where it was 6 feet thick every year a decade ago, now it was only 3 feet thick.

Dog teams were tied up throughout Old Crow, panting in the stifling August heat. More and more Gwich'in are returning to using dog sleds instead of snow mobiles to save both money and greenhouse gas emissions. But with climate change, frozen rivers and sufficient snow is no longer something that can be counted on: even in the heart of winter.

I hiked with a group up to Crow Mountain, stopping to graze on blueberries that grew plump and plentiful in great abundance. At the top, a large meteorological tower stood 60-feet tall, next to a small building housing its equipment. The Gwich'in were measuring the wind speed to see if they could replace some or all of their diesel-generated power with wind power. All throughout the world, wind turbines are being installed; but not nearly fast enough.

Chapter 9

Winds of the Future

Just North Dakota and South Dakota alone have enough potential wind energy for the United States four times over," said Chris Rose, who works to promote renewable wind energy in Alaska. It is just a matter of distributing that energy, and developing it in places around the country. He said with the exception of parts of the southeastern states, almost every other place in the country has viable wind resources.

"We should be developing those wind resources. We can produce energy that is relatively inexpensive and emits no carbon except that which is used for the manufacture of the turbines themselves and putting them into place. Why not error on the side of caution and do something you know will decrease emissions."

General Electric signed its second $1 billion+ wind turbine contract with Citizens Energy and the Navajo Nation to generate 500 megawatts of energy with up to 300 turbines. London is building the largest wind farm in the world off its coast and soon one third of its electricity will come from wind generators. Global wind capacity increased 27 percent in 2007.

"In Alaska we have a special incentive to develop renewable energy resources and that is primarily because we have 200+ remote native villages that rely on imported diesel fuel to produce their electricity," Rose said. "And they are paying 40-50 cents a kilowatt-hour, which is stifling their economic development and makes it real hard for them to live. So they have really been the forerunners in wind technology in this state because they have such an incentive to do it.

"There is a village of about 5,000, Kotzebue, that is on the coast in Northwest Alaska. It is the hub for all of the Northwest Arctic communities. In 1997, with the leadership of Brad Reeve, the general manager, they started to develop wind power again.

"In the early 80s, when the technology wasn't that good and the federal tax credits still existed, there was a lot of investment in wind power all over the country and especially in Alaska that failed because A) the technology was not that good and B) people were just getting the tax credits and they did not care if they operated or maintained them properly. So it left a bad taste in people's mouth about wind power.

"Back in the early 80s the cheapest you could produce wind power might have been 35 cents a kilowatt hour whereas now you can do it at three cents. So the technology has totally changed. Brad saw that in 1997 and got some money from DOE and put in a 50-kilowatt hour wind turbine that was successful and has since added them every year and now they have about 600 kilowatts of installed capacity there, and it is working great in a very harsh Arctic climate.

"It has been a stepping stone for some other villages: Wales, Gamble, Saint Paul, Port Heident and Pilot Point that already have installed wind power. The Alaska Rural energy plan is identifying over a hundred more villages that have wind potential. So the success of the Kotzebue project has made people think that it is possible to do it on a larger scale here in what we call the rail belt, which is the area between Homer and Fairbanks that is connected by electrical interties (the grid). It is essentially the same places that the Alaskan rail line goes, but not exactly, that is why they call it the rail belt.

"That whole area is connected (by the electric grid) and Chugach Electric, which is the largest utility in the state primarily serving Anchorage. They are very interested now in developing what could be a 50-100 megawatt wind farm off the coast of Anchorage on an Island called Fire Island, just a couple miles off the coast of Anchorage that could really jumpstart the harnessing of wind in Alaska. In April, 2008 the Alaskan legislature appropriated $25 million of a $250 million renewable energy bill to start the Fire Island Wind Farm.

"It would be right in the largest city in the state. If they did a hundred megawatts we would have roughly 20 percent of our load potentially served by a renewable source. That would really build enthusiasm for doing it in other places. If you look at the Department of Energy's renewable energy lab's wind atlas, they class wind in seven different classes. One is basically nonexistent; two is marginal; three is fair; four is good; five is excellent; six is outstanding; and seven is superb.

"If you look at the map of the United States, there are virtually no superb wind resources anywhere in the lower 48. And if you look at a map of Alaska, the entire Alaskan Peninsula, and the Aleutian Islands and most of the coast is superb. So we've got 98-99 percent of all of the best wind resources in the whole country here.

"One of the things that we foresee happening is we are ultimately going to have to move to a hydrogen fuel cell economy at some point. It is the only energy carrier that would allow the quadrupling of energy demand on this planet that has been predicted over the next hundred years. Nothing else is going to allow us to use as much energy as we are using. Fuel cells are probably the wave of the future.

"Most of the hydrogen made in the world now is made from stripping the carbon atoms off of the hydrogen atoms from methane or natural gas. So you've got two issues with that. Number one you don't have a huge net energy gain, because you use energy to do the steam reformation process but you also have to do something with that carbon. And if you can't sequester it you still have carbon emissions.

"What we would be looking for eventually is a way to produce hydrogen in a way that does not use fossil fuels or emit any greenhouse gases. We have a way to do that if we have cheap wind power, you can use the electricity to do electrolysis. Because otherwise electrolysis doesn't have a very big net energy gain either if you use fossil fuels to run the electric current through water and make hydrogen through electrolysis, you are not gaining anything. But if you use wind energy or solar or any other renewable to make the electricity to produce the hydrogen, then you are essentially creating a full green cycle of energy because the only byproducts of a fuel cell chemical reaction is electricity, a little bit of heat and potable water. So there are no emissions from fuel cells.

"That is what we see happening in Alaska: using some of our vast wind resources to produce hydrogen. There are already companies making fuel cell cars and fuel cell buses. In Reykjavik Iceland, the city bus drivers pull up to a Shell station that dispenses hydrogen into the fuel cell buses. And that hydrogen is being produced from electricity off Rechavick's grid that is all created by either geothermal or hydropower. They've got an electrolyzer in the Shell station, they do all of the electrolysis right there at the station. There are 80 fuel cell buses in Reykjavik now.

"Amsterdam just opened one last month. Hamburg and some other places are to open up the same thing. There are companies now that are making fuel cell cars and buses. One is New View. One of the exciting things about the Fire Island project near Anchorage is, if we could get that installed, we might be able to make some hydrogen from that wind power and have one of these same kinds of filling stations right in Anchorage. Then you could maybe convince the municipality of Anchorage to have some buses or fleet vehicles that run on hydrogen.

"Right now one of the big problems with hydrogen besides how you make it is how do you distribute it and store it. So these schemes of using the vast wind resources out on the Aleutian islands of Alaska don't make a lot of sense today because we can't economically transport that hydrogen from a place like Dutch Harbor to Tokyo. The cryogenic tankers and the things that might be used to transport liquid hydrogen or compressed hydrogen are just not economical yet. But if you can do the electrolysis right where you need it, then it makes a lot of sense.

"In Reykjavik, they have that electricity right there. If we can make power in Anchorage, and make some hydrogen from it, it could work. For the next twenty years or so, you are going to have to find places close to where you have the demand to make hydrogen work.

"We as a society have not wanted to put a price on environmental degradation very often. If it costs a hundred million dollars to move the village of Shishmaref because there is less ice on the coast, because it has been melting, that is something that we might be able to attribute to this climate change.

"If it costs 35 million dollars for the state of Alaska to repair roads because the permafrost is melting, that is the kind of thing that we understand. We don't understand or put a value on species extinction. If the polar bear goes extinct in a hundred years, what will that cost. I think some economists do analyze things like existence value. That is important in the analysis in Alaska because people who might not travel to Alaska still put a value on pristine wilderness existing."

Rose said economists have put a value on ecosystem services, which are being provided by nature at 37 trillion dollars a year. "That is equal to the price of the world domestic product each year, 37 trillion dollars," he said. "Those are ways to start putting prices on things like species extinction and cultural changes." While deep moral issues are raised in putting a price on nature -- the source of life -- it is critical for lawsuits to hold the coal, oil and auto industries accountable for climate change.

Climate change is a major cost that is not being factored in to the price of fossil fuels, Rose said. "Since there is almost a month less river and lake ice in Alaska, that is making a difference on cultures because most people will travel more in the wintertime in rural Alaska than the summer because there is ice, and you can go places, and you can visit. A lot of places in western Alaska that have water everywhere are really hard to travel in the summer. You've got to have a boat or a plane, you can't just go by snowmobile or dogsled like you can in the winter. So if there is a month less of ice, it means there is a month less of communication between people in some instances. That is a cultural problem.

"It is also a cultural problem if you can't eat the fish from the Yukon River or if you can't find the other foods that you would normally find because of ecosystem changes. So the externalities are not counted at all.

“These are externalities that are finally being measured, but the connection is not being made by most average folks. That is an important thing to consider. What happens if we can’t live the way we used to live. That is not a question that most people are considering. But it is a huge externality.

“The other thing about the prices, one of the things that is really important for wind energy development still is the 1.7 cents per kilowatt-hour production tax credit, the PTC the federal government gives to for-profit utility companies for producing wind. Even without the 1.7 cents per kilowatt-hour subsidy, we are still looking at very competitive wind projects in the lower 48. The PTC is essentially there because Congress feels that wind has to be given some equal footing because the other energy sectors get subsidies.”

Concern that the PTC would not be renewed was in part responsible for the slowdown of new wind energy production in the United States from 1635 megawatts in 2001 to 429 megawatts in 2002. Rose said Bush’s energy policy has been to further subsidize polluting fossil fuel industries instead of investing in solar or wind.

“The energy bill that failed (in 2003) had billions of dollars in subsidies for the oil and gas industries. When we are looking at policy it is important to consider who is getting what. It is amazing to me that the tax deduction that an average person gets if they get a Prius went down in 2004 from $2,000 to $1,500 but if I want to go out and buy a 6,000 pound Hummer that gets 8 miles per gallon, I would be able to write the entire $60,000 off. It is crazy.”

Taxpayers for Common Sense agreed, noting that purchasers can take $106,000 off their taxes of the $110,000 purchase price of the most expensive Hummer, and that businesses can also write off the purchase of SUVs from their taxes. The group issued a statement December 12, 2003 stating, "The tax exemption has become a misguided incentive for people to buy much larger vehicles than they need."

Rose continued, "Policies are really important. I am looking at what we can do in Alaska to provide some more policy incentives for renewables. One thing that we might be able to do up here is to get interested companies involved in what is going on in the bush because those little villages need much different kinds of wind turbines than the big utility sized ones that are being proposed for Fire Island or that are being used in the lower 48 in wind farms because these villages might only have a hundred people; they don't need a huge turbine.

"Some of the big turbines are large enough now that they can electrify three thousand homes with one turbine. Obviously you don't need that kind of thing in a small village. But the companies that might want to get involved and continue to make the smaller turbines might be able to market those same kinds of systems to the developing world where there are a lot of places that still don't have electricity. The whole idea of leapfrogging technology comes into play. There are lots of places in Africa that are never going to have phone lines, because they didn't have phones, and cell phones got invented, and now people are getting cell phones. They will never have telephone lines there.

"The same thing could happen with electricity if they could find a way to create electricity in their area with renewables. They would not necessarily have to be connected to the grid, which has all of its own downfalls as we saw last October (2003) with the power outages. The more you are connected, the more vulnerable you are in many ways. If you have what's called distributed generation where you can produce power more locally, you have more control over what is going on."

Eventually wind and solar on the grid would help to reduce blackouts, Rose said, but only after a massive increase in their usage. "We are so reliant on centrally generated electricity. The answer is yes (it would make us less vulnerable to blackouts) but it would take a long time.

"If we had a hundred percent wind in this country, we would still have to have some transmission lines, because the wind does not always blow all the time everywhere. So we still have to have some transmission lines, because if the wind wasn't blowing in Avaline, Texas one day, they would have to get their wind power from somewhere in Oklahoma maybe, because the wind does not always blow one hundred percent of the time.

"For instance in Denmark, they are up to where twenty percent of their electricity is being generated by wind right now. And what they are finding is the more places that they put the wind generators, the more stable their wind grid is becoming, because the wind might be blowing in place A but not place B.

With a more serious investment in wind energy, Rose said "it wouldn't take long for these industries to develop. There is all kind of information out there, and analysis and predictions of how much spending would do what. Even a relatively modest in the greater scheme of things investment of a hundred billion dollars over a period of years would be a huge boost to the renewable energy economy. A hundred billion dollars is not that much when you consider that we are spending four hundred billion dollars a year on our military right now.

"People have to accept the argument that a big portion of that is to protect Persian Gulf oil access. So why don't we take some of the money from our military budget and from oil and gas and put it toward subsidies and incentives for renewables, or hydrogen fuel cell research and we wouldn't have to protect our Persian Gulf oil. It really is not that much money in the greater scheme of things.

"We should pay at the pump what it really is costing our country. In Europe they are paying five dollars a gallon, and I would argue that here we are paying much closer to five dollars a gallon because much of those costs are downstream in higher taxes to support things like the military that protects our oil. Even though we are not paying five dollars at the pump, there's other hidden costs in our economy like the military budget that are attributable to our dependence on fossil fuels.

"There are a lot more jobs in the renewables and the efficiencies areas than there are in oil and gas. A lot of oil and gas are automated technologies and there aren't a lot of people involved. You can create a lot more jobs as well if you hire a bunch of people to retrofit lighting and tune up air conditioners and things like that than you do in just producing more coal or oil.

"I know there is a lot of job potential (in solar and wind). And they're sustainable jobs. That's the thing about a lot of these jobs is they will last forever, because the wind is going to keep blowing. You are going to have to keep putting new turbines up and maintain them. But the oil jobs and the gas jobs, whether it is in 20 years or 100 years, they will go away. One of the big topics of discussion up here is what are we going to do with our fiscal gap.

"I just got back from a two-day conference in Anchorage called Alaska 2020, which is a visionary kind of process of where are we going to be in 2020. One of the big issues that came up at this conference is we don't have a very diversified economy. Two-thirds of our economy up here is either Ted Stevens giving us money because he (was) the chair of appropriations, or oil revenue. Both of those things are going to end." Stevens' chair position did end in January, 2007, when the newly elected Democratic Congress took over. He is now without a job and convicted of seven felonies.

Rose continued, "We really need to look at diversification. And we have the ability to continue to be an energy state; we just have to use our brains around the idea of being a different kind of energy state."

While some have complained about altering natural viewsheds with wind generators, Rose said, "To me, it is a visible sign of progress that we would have our energy generated by renewable energy sources. And the fact is we don't see the emissions that are being created by the burning of fossil fuels for electricity so it is out of sight and out of mind. I think it is nice to see where your electricity comes from.

"I do think that there are places that are very special that we should consider the viewshed of. But for the most part, I don't have a problem with the viewshed issue. We have worked with the conservation community in Anchorage to develop support for the Fire Island project. In fact we even developed support for a smaller pilot project with wind energy that Chugach electric was considering in Chugach State Park, right in one of the most beautiful viewshed in the state in a fjord right south of Anchorage.

"We even got support from the conservation community to do that because we realized that the tradeoffs are so important. On Fire Island, it s not going to be a big issue because it is far enough off shore, nobody lives on it, and it is not going to be a viewshed issue. Even if it was a viewshed issue, I think people would be Ok with it because the conservation community here in Anchorage realizes how important it is to begin to wean ourselves of fossil fuels.

"You can see it and say 'Wow, look at that.' We were even considering having a Kiosk right there explaining what it is all about and get people excited instead of just assuming people are going to just like it. A lot of people have seen Altamont Pass in California which was one of the first big wind developments in our country. Those places were both sited poorly. First of all, they killed a lot of raptors, because they sited them on ridgelines where the raptors fly and they killed a lot of birds. Secondly, they have continued to kind of add haphazardly to what they started with.

"Even though they started with what would be considered today tiny little turbines, they've begun to add the really big ones and it is a very hodgepodge look. With the new wind farms that are being created today in Iowa and Minnesota and Oregon, they're just putting up these huge, sleek turbines. They all look the same. The blades are so big that they actually turn slower, and they make less noise.

"It is not the hodge-podge look that you get in Altamont Pass in California. They are all big, and there are fewer of them because they are much, much bigger. They will still kill birds. What is happening though is people are getting smarter about where to site them. If you put a new wind turbine in a bad place, it is still going to kill a lot of birds.

"What they are doing is doing more pre-construction studies to see where the birds are, and then doing more monitoring afterwards. They are doing a lot less damage to the avian population just because of better planning. And you also have to put that into perspective. There are an estimated one billion birds killed each year from manmade structures. One billion, and 80 million by house cats. And about 40-60,000 birds are killed a year by wind turbines. So it is a really tiny amount of birds."

Indeed this number pales in comparison to the number of birds that are being killed and are expected to die because of climate change. A 2003 study found that up to 37 percent of bird species could go extinct by 2050 because of climate change, unless we make major changes to reduce emissions. With the pace of warming so dramatically greater than just 5 years ago, more recent studies are now pointing to an even larger extinction rate if we continue with business as usual. Putting the issue into perspective, wind generators are actually saving the lives of exponentially more birds than they kill by reducing greenhouse gas emissions, though we still need to try very hard to minimize their direct impacts.

"What people are still concerned about, and rightly so are raptors, and endangered species. So when the Fish and Wildlife Service is helping people to site a wind farm, those are the things that they are looking for: raptors and endangered species.

"The other thing that has changed to make fewer bird kills, is now we have these tubular towers, instead of the lattice type towers. The lattice type tower attracted birds for nesting because they had a place to sit. So it is better to have the tubular towers where there is no place to nest or roost and hang out. That is another difference.

"The tip speed is still pretty fast. The rotor speed is slower. But way out there on the tip it is moving fast. They are also experimenting with different things like lights and colors and things like that. Lights are found to attract birds, at least certain color lights, so it appears that lights are not good. I think that most of the decrease in kills is figuring out what birds you've got before you build it so then you can design and locate it in areas where it doesn't affect them as much.

"Viewsheds and birds, those are the big issues with wind power. And I don't think they are that big, but they are issues. I think energy is the most important issue. It affects everything. It affects our environment, it affects our economy, it affects our geopolitical policy. So it is the most important issue in my mind for any politician at the national level to be focusing on."

Chapter 10

Bringing Testimonies from the Arctic to the Masses

With the dramatic impacts of climate change seen in the Arctic, the follies of our energy policies are all the more clear. Instead of pumping oil from the Arctic, we could be installing wind generators and solar panels, so that clean, renewable power could be generated much closer to where it is used. In remote places, such as Old Crow, wind generators can provide energy independence in a place where diesel fuel is currently flown in to fuel generators.

Any impacts of wind generators pale in comparison to the devastation seen by mining or drilling for fossil fuels, and the climate chaos that is caused by burning it. The native peoples are on the front line of the climate crisis, and their message must be heard, while we have a chance to make a difference.

"The weather is really unpredictable," said Gwich'in Sarah James. "There is extreme change. We don't have a healthy four seasons anymore. The cycle of the season is unpredictable, and that affects the animals that depend on those things. We have long, mild winters and all of a sudden it will change to real cold weather and that will kill animals. When there is too much extreme change, they can't adapt to that.

"Sometimes in the winters it gets really warm, and that is not good for animals either. It makes the river flow, and it makes it hard for us to get wood and hunt." Wood is easier to harvest when the ground is frozen because transporting it is easier on a solid surface. "The ice breakup is unpredictable, too. Sometimes, it breaks up early, or stays too long. It used to be much more predictable."

I had stayed with Sarah in Arctic Village in 1993 during my second expedition, just before she took me across the East Fork of the Chandalar River in a boat, and I climbed over Misty Mountain. Even back in 1993, and during my 90-day Arctic Quest in 1991, I found significant signs of climate change in the Arctic.

Working with organizations such as Greenpeace, I began asking for action on climate change in 1989. Then, few understood the threat global warming presented. I lobbied for strong action during the United Nations talks in New York, and during a week-long fast in front of the White House in 1992. In 1993 I went on a five-week speaking tour of Japan, in part to speak about the threat of global warming and the need for a strong international agreement to cut emissions.

It was after working for a year with Ozone Action, which merged with Greenpeace USA in 2000, that I realized how serious the issue was. I spent six weeks in The Hague, Netherlands at the United Nations Convention on Climate Change, working with Greenpeace to show support from the U.S. for strong mandatory cuts in emissions. I was impressed with Greenpeace, as they combined banner drops with suit and tie negotiations very effectively and professionally.

Arctic Screaming

Greenpeace activists chained themselves to a coal barge, shutting down Rotterdam Harbor. I witnessed Greenpeace activists chain themselves to the truck they stopped in the middle of the road in front of the UN building. They took the tires off and poured cement onto the chained in volunteers. Meanwhile a siren blared: an effort to sound the alarm on this issue that we should have made number one back then.

I cheered support for the brave souls willing to risk death to get the message out on this critical issue. I also passed out fliers explaining the purpose of the demonstration, along with dozens of other Greenpeace volunteers. One of the facts that startled me the most was how 42 percent of the polar ice had melted in the prior 40 years. Having trekked through the Arctic Wilderness twice, I began looking into it in much greater depth.

Two years prior, Greenpeace sent researchers up to the Arctic to get witnesses to the change that scientists knew was occurring, and publicized those findings. Greenpeace "wanted to work to educate Americans on Global Warming and raise the level of concern," said Dan Ritzman, who worked on the project. "The second component was we wanted to do something directly to combat climate change. Greenpeace was unique in that it is international, and they did the science to prove that we are getting to the critical point of no return. We cannot burn a quarter of our known reserves. So there is no point in finding new reserves. We should use that effort to find renewable energy sources.

"Greenpeace's focus was on the western Arctic because the Arctic is warming faster than any other area. One way to point out global warming is to get testimonies of the native peoples who have a closer connection to the land. We thought they could relay in their lifetime changes in the land that they have seen.

"For the people of coastal Alaska, subsistence is still an important part of their life. They talked about changes in the pack ice. It would come in later and go out earlier. There were more violent storms. There were earlier springs, changes in vegetation, more shrubbery and trees and occasional anecdotal evidence of new species in the ocean and land.

"We went to eight villages in total. In each village we would have a general meeting—usually 20-30 people. Often we would focus on one or two people who were more articulate or had more to say. For us it was a pretty eye-opening experience. We believed climate change was happening. It was in the universal nature of what people were talking about. There were so many similar bits of anecdotal evidence it was hard to ignore them. The bumblebees are bigger. The ice is thinner, they would say. After you've gone to five or six places and the same similarities are said, it is hard to ignore."

Writer Yereth Rosen reported on April 21, 2004 that "ANYONE who doubts the gravity of global warming should ask Alaska's Eskimo, Indian and Aleut elders about the dramatic changes to their land and the animals on which they depend." Indigenous peoples of Alaska report that the salmon are increasingly susceptible to warm-water parasites and have lesions and odd behavior. Native peoples said salmon and moose meat have odd tastes and the marrows of moose bones are weirdly runny.

Three warm summers in a row have dried wells and forced the native village of Nanwalek in southern Alaska (population 214) to ship in bottled water and ration it, Rosen wrote.

Deborah Williams spoke about the plight of native peoples in Alaska because of climate change.

"They rely so much on hunting and fishing and gathering. When it becomes harder or even impossible to hunt because you don't have the ice to hunt on; when your knowledge about how to hunt is no longer relevant because things have changed more dramatically; when people die more often hunting because they fall through the thin ice, or broken off ice; when the gathering of berries becomes more difficult because the conditions for berries in Alaska is much worse, it is so disruptive of cultures. Sterling believes that climate change will potentially lead to the destruction of the fabric of many Alaska native cultures and Siberian Yupik will be one of them."

Rosen reported that "Indigenous residents of the far north are finding it increasingly difficult to explain the natural world to younger generations." Larry Merculieff, an Aleut leader from the Pribilof Islands in the Bering Sea said, "As species go down, the levels of connection between older and younger generations go down along with that."

Charles Wohlforth recently published, The Whale and the Supercomputer: On the Northern Front of Climate Change (2004), about perceptions of climate change in the Arctic. In an interview, he said, "I think that with the warm weather that we have had in the last few years there has been a pretty broad awareness that something in the climate has changed. And I think that is pretty universally accepted."

A 2006 poll found that 81 percent of Alaskans are convinced that global warming is happening (Leiserowitz & Craciun). People were asked if global warming is already causing or accelerating the following: for the loss of sea ice 83 percent agreed; 82 percent for the melting of permafrost; 74 percent for erosion, and 72 percent for forest fires. Seventy one percent of Alaskans call global warming a serious threat to the people in Alaska; 76 percent said global warming is a serious threat to plants and animals and 81 percent of Alaskans favor government regulations on greenhouse gas emissions for power plants.

A February 2008 poll conducted for the nonpartisan Civil Society Institute found that 86 percent of Americans agree that "America should commit to a five-year moratorium on new coal-fired power plants and instead focus on aggressive expansion of wind, solar and other renewable energy sources."

What we need is massive, professional, civil action as was seen in the civil rights movement so that our government leaders act on the will of the people, rather than that of the coal, oil and auto industries. I have seen massive change when a critical mass of citizens get involved. Future generations depend upon us getting active politically and making enough of a difference now more than ever, while we have a chance of slowing things down to keep Earth livable.

Native Villages are being severely impacted by the receding pack ice, Wohlforth said. "The sea ice is much less reliable as a platform for hunting. This has several impacts. One is the increased danger because of ice breaking off and people floating out in the ocean on the ice." He said it is difficult to find ice strong enough to haul a whale upon. The period of safe operations in the spring is shorter and more interrupted by dangerous incidents, which makes hunting less successful. In the fall the retreat of the sea ice has caused much larger waves to be present there along the Arctic coast, which means that hunting in boats is more dangerous because the waves are rougher. And it also dramatically increases the erosion on the shoreline.

"If you go closer to the southern edge of the ice pack around St. Lawrence Island, you have a situation where you have people that traditionally hunted on ice much of the year, but the ice just simply did not show up as much of the time as before. And this is true around Barrow too. There is less of the good time of hunting when there is plenty of light and when the ice is available."

During nearly three months of the Arctic winter, the sun never rises. It is in the fall and spring where native peoples had the overlap of sunlight and ice. But these are the times when the ice has decreased the most, cutting back on that critical time when hunting is the easiest.

"It was very clear to everyone that something very dramatic is happening," Wohlforth said. "It has already affected their way of life and how they pursue their subsistence. Having kind of a holistic or dynamic view of nature in some ways is a much better way to see and understand climate change and how it is happening.

This contrasts with the sort of snapshot version or the computer climate-modeling version that the scientists produce. There is more certainty in actually living out nature and experiencing it and seeing how it changes through time as a native elder than as the scientists who plug a lot of data points into the computer to get an answer."

The value of traditional knowledge is not individualistic. It is based on a lot of people in the community who have worked together and developed a joint body of knowledge or a consensus on what is happening. A lot of the elders have said the weather has become less predictable. It is more of a matter of patterns. There is a different size of storms, and they tend to be larger. There are warmer temperatures in the summer which make it more difficult to preserve meat that they have hunted out on the tundra."

Warmer temperatures in the Beaufort Sea has made fishing more difficult because the fish no longer preserve like they used to when it was colder, reported Susie Akootchook of Kaktovik. "The ocean has warmed up tremendously. So much so that the fishermen have to check their nets out more frequently so the fish won't spoil due to the temperature of the water."

The Arctic Research Consortium of the United States (ARCUS) published "The Earth is Faster Now: Indigenous Observations of Arctic Environmental Change" in 2002 about indigenous peoples in the Arctic and their observations of how the environment and climate has changed.

Wohlforth said, "When they say the Earth is faster now, they say that the Earth's systems move through quicker and are less predictable, and that fits in pretty well with the observations. What the elders are saying is the weather systems are moving through faster, they are less predictable and they are more violent. And after that observation was made by the elders there were some meteorologists who went out and studied the weather patterns and said 'by golly that is true.' The storms that are coming through are coming in on a more frequent rhythm than they normally did. That is something where the elders noticed something very subtle about the patterns of weather coming through and then the scientists went back and were able to confirm it in their own perspective. I don't think any of it is a big secret anymore."

Global warming has become a common term in the vocabulary of the people in Alaska, Wohlforth said. "Last winter (2002-2003) was the warmest winter that we had up here. There were phenomenon seen all over the state that were rarely ever seen before. Rivers never freezing, lack of snow cover, animals not hibernating that normally hibernate, insect populations exploding because there was not enough cold to kill them off during the winter: there are just huge numbers of unusual occurrences due to very warm winter conditions. We had several warm winters." Wohlforth said three winters in a row from 2001-2003 were all quite warm and there were a lot of impacts from it. The author warned that "the biological ranges at the furthest north will presumably disappear."

Chapter 11

Boating the Porcupine River from Canada to Alaska

From Old Crow, our plan was to boat down the Porcupine River with two Gwich'in guides. We would visit their ramparts camp, camp overnight downriver from there, then boat into the United States and over to the Gwich'in village of Fort Yukon. Our Gwich'in guide, Dennis, warned that water levels were going down, and that they might damage their outboard motors. While I would have much preferred to kayak, canoe or raft, the group was on a tight schedule, and could not afford the time.

We loaded up our gear, and launched on the long journey. We were immediately engrossed in amazing scenery with evergreen trees silhouetted against the sky along the banks and mountains. Amazing rock formations jutted up along the banks: rarely seen monuments to nature's beauty when she is left in peace. But even here, in one of the remotest places on the planet, the reach of our species through climate change was clear to the trained eye.

Erosion ate under the banks of vegetation, which draped over the edge of the riverbank like a carpet. Permafrost had once held in the riverbanks, being near impervious to erosion. The rapid warming has crumbled the formerly frozen earth and floods have swept that silt downstream. Deep pools are filling in with that excess silt, harming fish habitat and the whole food chain.

Climate change likely contributed to the smoke in the air as well, that created a haze and made me wonder how safe it was to just breathe. In addition to the massive death of millions of acres of forest in southern Alaska, trees were dying in the interior of the state as well, and climate change is the most likely culprit.

Rather than crystal clear skies with visibility in the scores of miles as I had seen in the Arctic in 1991 and 1993, a smoky haze hung over the Arctic, getting thicker or thinner based on the direction of the wind. All over our planet, climate change is killing forests and exponentially increasing forest fires.

In Montana, average March temperatures have risen five degrees Fahrenheit in the last 50 years, and average snowfall in Missoula has decreased 22 inches in the same time period. With warmer temperatures and less moisture, Montana experienced five 100-year forest fire years in the last 15 years.

Major forest fires increased four-fold when the time period 1970-1986 was compared to 1986-2005, scientists reported in the journal Science July, 2006. That report found that the length of the active wildfire season increased by 78 days and the average burn duration of large fires grew from 7.5 to 37.1 days.

The cause of this increase is the mountain snow pack melting 1-4 weeks earlier, and spring and summer temperatures that are .9 degree Celsius higher. This leads to much drier, fire-prone forests. But the fires pale in comparison to those seen in Alaska in recent years.

Chapter 12

Smoky haze from Crow Mountain

A record 6.7 million acres of forest burned in 2004. Millions of acres of spruce forest are dead across southern Alaska, but northern Alaska has not been spared. Impacts have spread through central and northern Alaska, with higher temperatures contributing to drier forests. Forest ecologists warn of much more devastation as the Arctic continues to warm.

Scientist Glenn Juday reported that he has found a temperature, 16 degrees Celsius, at which when achieved and exceeded, tree growth declines rapidly. This temperature had previously been quite rare, but with climate change has become more common at the expense of the forest.

Juday has been studying the reasons behind the massive die-out of trees in southern Alaska, as well as what the future will be for the northern landscape. "I am part of the long-term ecological research group at Bonanza Creek. I have been doing long-term ecological research in general for many years," he said. "I use the tree-ring technique to look at these things.

"In looking at the growth of trees in the boreal forest region of Alaska, we know that the trees in particularly sensitive locations, such as trees in the tree line, are under a real close control by the climate. In good years they grow a whole lot more than poor years. This relationship is so good that you can actually use the tree rings to reconstruct the past climates. The classic way of doing that is to look at the ring width, to see how much growth there was.

"But there are other techniques, like carbon 13 isotope content of the wood year by year. That goes up and down with changes in the environment. And finally the density of the wood as measured by X-Ray means. So we've used all three of these techniques. We looked at several white spruce on our most productive upland sites in the boreal region.

"We found that contrary to what we expected, the trees grew the least in the warmest years, and they grew the most in the coolest summers. That just moves exactly the opposite of what people had expected.

"We used that data to reconstruct past climates," Juday said. "When we tried that much further south and lower in elevation, and the really productive boreal white spruce stands that are in and around central Alaska, we found just the opposite compared to the tree-line trees. And not only that, but we have climate records from the Fairbanks Research Station, from about 1906 onward, so we have most of the twentieth century. When we looked, we found that the last couple of decades in the century were the warmest summers and sustained. And that was the period where the trees grew the worst, and by far, quite a bit.

"Now we have begun to extend that research. I just had a PhD Student, Martin Vilkey who put together a big data set on the Brooks Range, along with some cores from the Alaska Range, and another

Master's student Herald Zault, who did work in the Alaska Range. We had a massive sample, with more than 1,400 trees cored. We had 13 sites, and two mountain ranges so it is really reliable results."

"When the temperature was colder than 16 degrees Celsius at Fairbanks, it got some production... But boy after 16 degrees their growth went way down, a real tight relationship. And that indicates that there is a threshold affect. So as it gets warmer during these critical time periods for their growth, it really subtracts from that growth.

"We split the twentieth century in half, and showed there were significant predicted relationships before 1950, but after 1950, they were real strong. The difference was simply that after 1950, we had a lot of these years of warm summers. It is really reducing the growth of the trees on our sites.

"We went one step further and extended that work to black spruce and paper birch, and found both had negative responders. We found a lot of negative responders to paper birch. We are kind of filling out the picture now by region and species and responder types. And we are finding that a substantial number of trees in our forest are being reduced in the amount of growth they can achieve because of warm temperatures. The warm temperatures have become more common. There is a threshold temperature. Where it is warmer than that, growth is really reduced.

"Since that is a straight-line relationship, we've taken the climate change scenarios that have been run for the 21st century;

five different climate scenarios. We used those scenarios to see what would happen. Because there was a straight line relationship, we found that several of the negative responding species site types during that 21st century scenario period got warm enough that the relationship showed zero growth. With these scenarios, then these species will not be able to grow anymore during the warmer times.

"The second issue is, wait a minute. If you stress some of your house plants, for example if you don't water them, they can get so weak that you come back on vacation, and you water them but they go ahead and die. It is the same thing in the wild. If you stress a plant badly enough, two things happen. It can reach a pivotal stage short of the zero line that I am talking about, where the temperatures are so high that growth is zero according to the predictions. Clearly the plant will probably kick the bucket sometime before you get there."

Because of increased emissions and understanding of feedbacks, models today are even worse than those Juday used. We could see signs of past fires as we motored down the river.

Though a pleasant journey, I kept wishing that I were paddling the river quietly with a kayak or canoe. But I knew that this group did not have the time for such a venture. We entered the Yukon Flats region. Thousands of lakes dot this vast wetland region: the destination for tens of millions of ducks and other birds.

As the Porcupine River entered the Yukon, I was stunned by the mile-wide size of the swift current of murky water. It was a short ways around the bend of the delta to the shore of the Yukon, and the Gwich'in village of Fort Yukon. We camped on the shore of the mighty river and enjoyed some beers.

Moses Lord, a Gwich'in from Fort Yukon said, "I remember growing up as a child, the snow used to stay around until the first of June, ice used to break up in mid June. Now it breaks up the first week of May. In the last few summers, it's cooler with lots of rain and warmer winters. We notice some lakes are draining out, while other lakes have more water."

Climate change also has polar bear wandering farther inland than ever as they are unable reach pack ice in the Arctic Ocean now hundreds of miles from shore when it once rarely exceeded 15 miles. A polar bear, unable to get out onto the pack ice, which was once accessible year-round from the coast of Alaska, was killed 250 miles from the coast, near Fort Yukon in March, 2008. The previous record for a polar bear was 125 miles inland.

From Fort Yukon, we flew to Venetie in an 8 passenger double-propped plane. Venetie had installed the northernmost solar power system in the world, to keep their meat frozen during the summer (necessary now because of climate change). We next buzzed down the gravel runway, and lifted up over the small village, before heading on to Arctic Village. Wetlands, lakes and patches of young forest extended to the horizon in all directions: a sea of wilderness.

This was my third visit to Arctic Village, so Harvard sent me with the first flight. In the village, we heard about plans for the northernmost solar power array in the world. I returned to Arctic Village for more than a week after leaving the wilderness, to hike over to Old John Lake. When I returned, I witnessed the completion of the northernmost permanent solar tracking array in the world. These are among the poorest peoples in our country. If they can afford solar power, anyone can.

Chapter 13
Solar Powering the World

Having helped to build a home that ran entirely off solar and wind energy near Athens, Ohio, where we were told it was too cloudy and not windy enough, and seeing it easily power the living of 11 people who lived in the home, I am convinced that both offer the solution to our energy needs. It was actually cheaper to build a solar and wind home than to pay the cost of extending the power lines to the house. And that was in 1994.

Like computers, the costs are plummeting for solar and wind energy, making them cost-competitive with fossil fuel energy production, not even taking into account the externality pollution costs that make renewable energy vastly cheaper than their fossil competitors in the long run.

With a battery bank, and the constancy of solar and wind energy, our lights never went out. Meanwhile, our neighbors, who ran on coal-generated electricity that fed into the grid they were on, had regular power outages. The coal industry lied with their propaganda that "coal keeps the lights on," plastered on billboards and media advertisements. Coal-fired power plants emit one-third of human caused greenhouse gas emissions in the United States.

The truth is that our lights were powered cheaper with solar and wind even in the skewed economics of our day that falsely favors fossil fuels (Ohio gives $3 to coal companies for every ton of coal mined in the state). The state also spends tens of millions of dollars directly subsidizing coal mining, such as building their rail lines, every year. It was our solar and wind power system that kept the lights on. Meanwhile, the coal industry lights regularly failed, while contributing to the death of the planet through climate change, acid rain, mercury pollution, particulate pollution, acid mine drainage, strip mining, mountaintop removal devastation and more.

The amount of solar energy available on earth is so much that in one minute, the equivalent energy is received by Earth as the world uses in an entire year. But different parts of the world have different amounts of solar energy based on the season, elevation and cloud cover.

Just a small fraction of our south facing rooftops is all of the energy that we need for all of our electricity.

A new study revealed by Princeton University in September, 2003 found that they could produce organic solar energy cells, as opposed to the inorganic silicon cells now widely used. While the efficiency would not be as great for these cells, the costs would be so vastly cheaper as to make the overall wattage far cheaper. They could be made in many colors, even transparent sheets that could be used as tinting for windows while producing energy. This multi-coloring and ultra-thin design means it could easily be incorporated into highly aesthetic and diverse architectural designs.

A British Petroleum study in 1997 found that an investment of $550 million into a 500-megawatt solar photovoltaic factory would lower the production cost of solar panels by 80 percent. BP found that using existing technology in 1997, a large-scale factory can be built with no barriers to reach 500 megawatt per year production of photovoltaic modules using crystalline silicon. That is less than 750 times smaller than the annual U.S. military budget. And, improved technology since 1997 has made the cost even lower.

The San Francisco Chronicle reported that solar energy power production is growing 30 percent a year. The fastest growing sector in the industry is installing solar photovoltaic rooftop arrays on businesses, which often recoup the costs in four to eight years. Systems today are half the cost of systems seven years ago, reported Daniel Shugar, president of PowerLight, a Berkeley, California firm that installs solar systems. Every three years the solar energy industry doubles, and with that doubling of volume, prices are reduced 18 percent. Just a thousandth of the U.S. Defense budget put into producing solar power would double the growth of the industry.

Today, Japan, Germany and California lead in the global solar market, with 40 percent of solar production now in Japan, 20 percent in Germany and 12 percent in the United States. The industry is waiting for a Henry Ford of the solar energy business to streamline and reduce the costs even further. U.S. Department of Energy spokesperson David Garman said that the DOE is committed to reducing the manufacturing costs of solar photovoltaic cell modules to $3 per watt by 2010, and $1.50 per watt by 2020.

A partnership of Hong Kong and Australian companies announced in March, 2008 plans for the world's largest solar power station, expected to generate 154 megawatts, in northern Victoria, Australia.

While President Jimmy Carter put solar panels on the White House, President Ronald Reagan took them off. "President" George Bush Jr. put solar panels on a shed in 2003 in a token gesture of eco-consciousness that was only window dressing for an administration with the most anti-environmental record ever. In policy, Bush worked to remove subsidies from solar and wind energy, and put subsidies on the fossil fuel industry, which funded his "election" campaign.

Collecting methane from landfills, sewage treatment plants, livestock waste and coal mines and using it to generate both electricity and heat is a critical means to both massively reduce greenhouse gas emissions and generate renewable energy. This is because methane is 23 times more potent than carbon dioxide as a greenhouse gas. It also has great potential to generate energy. Because it can be stored, it can be a ready on demand fuel when winds are slow for wind farms or it is nighttime with solar panels.

Utilities are installing large battery packs to avoid the need for peak demand plants. A one megawatt battery was recently installed in West Virginia, delaying the need for a substation. This can also help to store energy as we transition to more solar and wind power.

The first wave-powered boat sailed from Hawaii to Japan in the summer of 2008. Wave and tidal energy both have great potential.

"If you have more renewables, you will create more jobs," Rose said. He said it would also decrease our dependence on foreign oil and improve relations with other countries. "We know that we are going to be reducing our output of greenhouse gases which would have a huge effect on our civilization ultimately."

Individuals can take actions in their own lives to reduce greenhouse gas emissions, particularly in the United States where we use so much fossil-fuel energy, Rose said. "Buy a hybrid car. Do more carpooling, do more walking. Replace lights with energy efficient lighting. Replace appliances like air conditioners and refrigerators and clothes driers with more efficient models. Educate yourself about what is going on around the world both as a result of emissions of greenhouse gases but also about the solutions. Urge utility companies to use renewable energy sources to generate electricity.

"People can demand from their local utility that they have the opportunity to buy some portion of green power. So there are a lot of things that individuals can do. We in the United States especially need to consider how much energy we are using. We take it for granted when we get on a plane or drive a car. But we are not really connecting that use of fossil fuel to what is going on with climate change. I think if we all educate ourselves a little more, we wouldn't want to do some of the things that we're doing and we'd all want to do the right thing.

"We all have an inner instinct to conserve and to do the right thing, but we aren't making the connection. We aren't using fossil fuels an armload at a time like when we are filling our wood stove with the trees that we cut up and sawed up. So we don't have that connection to the fuel that we're using. We don't see the tankers that supply our gas. We don't see the gas line going into our house. So it is out of sight, out of mind. I think we need to find more ways to develop the measuring eye that we have with simple fuels like wood.

"There is a sign in Manhattan that displays our national debt continuously. Why can't we have a sign that displays how much fossil fuel we are using? Why can't the little monitors in the hybrid car that tells you your real-time mileage (the miles per gallon as it varies during your driving), why can't those be standard equipment in all cars? If you are only getting eight miles per gallon, it is right in your face. Those are the kinds of things I think we need to put two and two together."

We know the solutions. We have the technology to do it. It is imperative that we realize our power as citizens to make a difference and change our government and industries to reduce greenhouse gas emissions. The fossil fuel industry, with its billions of dollars and entrenched desire to keep the status quo of destruction, has a clear advantage. But we cannot let the difficulty of our task get in the way of our action. There are millions of people around the world deeply concerned about climate change, and dedicated to making a difference.

The seriousness of this issue begs every citizen today to get active politically to institute the needed changes in government policy that would reduce our emissions of greenhouse gases instead of increasing them. We must demand that our policymakers invest in fuel-efficient trains, mass transit and bicycle lanes instead of polluting airplanes and highways, and mandate efficiency in cars, trucks and appliances. We must demand that subsidies for fossil fuels be replaced with subsidies for solar and wind energy.

The fate of the world is in our hands: the people in this pivotal time. It is up to us to make the future a living one. That means changing our present course, dramatically. It means changing our priorities. We need an army of activists dedicated to nonviolent social change through political involvement on a daily basis. It is up to us to be the difference, if the diversity of life as we know it is to survive.

Chapter 14

Movin' on up: Animals and Plants Migrate North

Talking to Gwich'in leaders, all spoke of global warming. They reported seeing new animals that were moving in from the south: such as black bear, robin and beaver. As the climate warms, the northern limit of animal and plant specie in the northern hemisphere moves farther north. As specie move into areas where previously they have not been, they are creating a massive change in the ecology, with predictions of the end of the tundra in Alaska.

"We see evidence of exotic species—animals that have not been seen before in these regions," said climatologist Guntner Weller. "The natives tell us that they see bird species they have no name for because they have never seen them before. This includes big land mammals. In Seward Peninsula, we see beavers. We see moose on the North Slope (of the Brooks Mountain Range). We see bigger stands of willows."

Arctic Village residents reported that beavers were damming creeks and stopping fish migrations.

"As the trees and bushes march north over what was once tundra, so do beavers, and they are damming new rivers and lakes, to the detriment of water quality and, possibly, salmon eggs," reported Yereth Rosen on April 21, 2004.

Yellow jacket populations are up tenfold in Fairbanks in 2008 and the first two sting related deaths were reported in 2006. Short eared owl and varied thrush were recorded wintering in Anchorage for the first time in February, 2008.

A $6 million documentary series on global warming's impact on the Arctic, Arctic Mission by writer-producer Jean Lemire filmed the new arrival of robins, ants, grizzly bears and moose; they have been seen in the Arctic for the first time recently because of global warming. The program ran in Canada on David Suzuki's The Nature of Things. Suzuki said that Inuit children have been afraid of dragonflies, which they had not seen before.

Gwich'in Moses Lord of Fort Yukon also reported seeing different animals. "There are a few different birds around. There were a few cougars spotted—at least three times. Black cougar were spotted (which had not previously been seen that far north). We noticed the caribou are confused. They go in different groups. They are not all together (in a single group) like a long time ago."

Arctic Village Elder Sarah James said climate change is reducing berry harvests. "During the summertime, sometimes cold spells will come and that is not good for berries, or the animals. One year there were no berries. Bear got really aggressive because they did not have the berries for food. And that goes for all the animals that depend on the berries. When that happens, they get attracted to human trash, and they become a nuisance for the village.

"Arctic Village—that used to be the tree line. We used to be nomadic. The creek was called the creek with high banks. It was a popular stopover because that used to be one of the favorite places to set fish traps. That was the tree line, but now the tree line is past the Brooks Range. The tree line is on the North side of the Brooks Range." Arctic Village is on the south side of the Brooks Mountains.

"We never used to have cottonwood. Now we have miniature cottonwood. We never used to have beaver or black bear. The last few years we have seen black bear. The black bear live with wood. Woodchuck and woodpecker came in— we never used to see them. Hawks and different kinds of birds came in. It was mainly open country before but there is more vegetation coming in." The lake by the airport in which I had caught a large northern pike during my first visit to Arctic Village was pretty well drained.

So much vegetation has grown into the creeks that it has obstructed the travel of fish, James said. "It blocks the fish in the creeks."

An increase in beaver populations and their moving north into areas they previously have not been have also been reported to block fish from reaching their spawning grounds. Paul Erhart, from Tanana said, "We now have more beaver than ever in this slough because of warm winters that give the beaver the most favorable conditions to survive. (Beaver) then proceed to dam and tier off the sloughs so resident species of fish, which again provides the Indians with a very viable source of food, cannot reach their spawning ground to provide the next generation of food for the Indians of the Interior."

Red fox have moved up to 600 miles north and established themselves in areas formerly dominated by the Arctic Fox, the Pew

Center on Global Climate Change reported in Novermber, 2004. While the Arctic Fox is better adapted to cold, in direct fights between red and Arctic fox, red fox will prevail. The red fox is not as adapted to the cold as the Arctic Fox, with longer ears and limbs that lose more heat. But as the climate has warmed so drastically in the Arctic, the red fox has moved far north. Scientists say the two species cannot co-exist.

On Baffin Island in far northeast Canada, the red fox has moved 600 miles north in 30 years. As the red fox migrated north, the Arctic fox retreated north, either due to warming, competition with the red fox, or both, the Pew Center reported. With the projection of a continuation of warming trends, the range of the Arctic fox will be further reduced. Scientists found that the red fox moved farther north in areas that warmed more. The Pew Center report found that in the Arctic, most regions have experienced a temperature increase of 4 to 7 degrees in the last 50 years.

"We are finding tuna where they have never been in Alaska," Deborah Williams said. "Species are now showing up that were not here before because it is warmer. The shrub line has dramatically marched north.

"The reason that Alaska was the place where it was predicted to warm the most, and has warmed the most, is that in the lower 48, as the temperature goes from 69 to 70 or 69 to 72, well it is warmer, but it is not a profound tipping point. In Alaska, when the temperature goes from 31 to 33, and you go from a temperature at which you create ice and snow to a temperature where you create rain and brown, that has a profound impact and accelerating impacts.

It rained while we were in Arctic Village. My tent leaked, and I was constantly drying clothes. We gathered around a fire, cooking and eating. I walked away from the village, toward a lake I had fished during my first visit to the Gwich'in territory in 1991, with Harvard, a church volunteer and a student. We came upon a startling sight. "Watch out," I said, pointing to a massive hole in the ground. "Look at that!"

Chapter 15

The Melting Permafrost

The ground had collapsed, about 30-feet into the earth. Two hundred miles above the Arctic Circle, just outside Arctic Village, a large chunk of ice under the tundra had melted, and the tundra above fell into the hole. I had been to the spot in 1991, and 1993. The collapse had just happened earlier in this, the summer of 2004.

In the Arctic, the land is supported by the permafrost—land which for millennia has been frozen year-round from a foot to 2,000 or more feet deep. The permafrost holds the lakes and streams in place. As it melts, lakes have drained and streams have eroded. All human structures in the two-thirds of Alaska that has permafrost depend upon the frozen ground for support.

The permafrost is very important to the ecology of the Arctic, because much of it only receives about eight inches of precipitation a year, about as much as a desert. The frozen ground holds that moisture up near the surface, and as the top layer thaws out every year, places like the coastal plain of the Arctic National Wildlife Refuge form some of the largest wetlands in the world, providing critical habitat for millions of birds from around the world, more than a hundred thousand caribou, musk-ox, bear, wolf and much more.

But global warming is turning the permafrost into a soft, slurry-like material that can trigger subsidence (the collapse of surface lands) and damage to buildings, roads, trees, power-lines, pipelines and more. University of Alaska at Fairbanks studies indicate that a change in permafrost temperature from minus four degrees Centigrade to minus one degree Centigrade decreases the load capacity of permafrost by as much as 70 percent. All throughout the Arctic homes and buildings are being damaged by the warming, with cracks and fractures as the houses settle into the soft ground.

The melting has also caused severe erosion of river, stream, lake and ocean shorelines throughout the Arctic. It has dried out many of the wetlands, destroying this critical habitat.

"In the last 10 years, we've seen a lot of changes," said Gwich'in Aaron Tritt in an interview in Arctic Village, the summer of 2004. "The permafrost is melting. The lakes are drying out. The tundra is collapsing. It used to be we went out hunting and it was noisy with all the wildlife and birds. Now it's quiet. We have told them what we can tell them. Even the Inuit know climate change is happening. If we see that evidence (of climate change) in the last 10 years, what are we going to see in the next 15-20 years? Twenty years is not that long."

Compounding the situation, the melting permafrost itself releases greenhouse gases that further warm the planet. The frozen earth has locked up a teraton of carbon for thousands of years. Svein Tgveitdal, director of the United Nations Environmental Program Center in Norway reported, "There is now evidence that this is no longer the case, and the permafrost in some areas is starting to give back its carbon. This could accelerate the greenhouse effect."

An estimated 14 percent of the world's carbon is stored in Arctic lands. As the permafrost melts, bacteria break down the organic material, releasing the ancient carbon in the form of methane or carbon dioxide: both greenhouse gases.

"As the climate begins to warm, the permafrost thaws, and with it comes pretty dramatic changes in the ecosystem," said climatologist Guntner Weller. "For example forests, especially if they have big ice masses, these melt, big holes develop and trees fall into it. We are seeing a change from boreal forests to wetlands and eventually over long intervals from wetlands to grasslands. We are beginning to see that.

"Practically all the roads in the interior are built on permafrost. The increased warming and settling of the roads also has an economic impact. Most of the interior of Alaska is in this discontinuous permafrost. In the discontinuous zone, the permafrost is patchy, and only a few meters thick, and absent elsewhere. The discontinuous permafrost is very close to melting. Fairbanks has an average of 28 degrees F. Four degrees will melt it."

Pam Miller, with Arctic Connections said, "If there is permafrost under or around the lake that was frozen, holding in the water, it melts and they drain. When you are going from Arctic Village on the East Fork of the Chandalar River to the north, there is a place where a large piece of the bank has completely melted out. I've had people up there point that out to me. It looked like a pile of mud where there used to be a solid river bank."

Charles Wohlforth said, "Numerous people I talked to said that the active layer was deeper than it was formerly, that formerly you could dig down two feet in September and see ice. Now you can dig down two yards. People are talking about streams cutting through because of the melting of the permafrost." Previously, indigenous peoples dug into the permafrost to store meat through the summer in permanent iceboxes. But, Wohlforth said, "Natives report a change in their ice cellars and less ability to keep food frozen and things like that. Those are pretty common observations."

Traditionally, Inupiat peoples stored whale, caribou, seal, walrus, fish or other subsistence meat in the frozen ground. But, global warming is requiring the Inupiat to provide refrigeration with electricity and freezers when previously such a necessity would have been laughed at in the far northern region.

Birds are having a similar problem, as food that they try to store for the winter is now rotting with climate change. A 25-year study by Professor Dr. Thomas Waite found that birds are getting less food, and are suffering from food poisoning because of rotting stashes of winter food, that once stayed preserved in the colder winters of the past (Proceedings of the Royal Society, 2006).

"Today more and more, global warming is real in the Arctic," Sarah James said. "A lot of our lakes have been drying out around the villages and in the valley. It's because they were two bodies of water, the lake and the stream. But the permafrost melted and gave way and one of the lakes drained into the stream and it dried out. It displaced fish spawning and fish movement. The fish are dying out because of no water or less water."

James said the permafrost melting "changes our homes." Around Fairbanks, where James has also lived, she said the highways have a lot of patches in the concrete when it sinks in.

Dan Ritzman said the native peoples already knew about climate change before the scientists predicted it. "Western science was confirming evidence that these people knew. When asked, they knew it was getting harder and harder to feed their families on subsistence hunting because the weather was more severe, and there were fewer days in which they could hunt. Some had issues with lakes. The lakes would drain and they would catch fewer fish in the lakes."

The native peoples had gathered and hunted food for thousands of years, building a deep base of knowledge of how to survive. But now this subsistence is more difficult, and the climate changing so much that much of their knowledge seems irrelevant for current conditions.

"It is very real up here," Miller said about climate change. "One time I flew between the Gwich'in village of Venetie and Arctic Village, and one of the women on the plane looked down and said 'Oh my God, is that lake gone?' It was one of the larger lakes (that dried out). It was empty of water. There are lakes that are drying up very fast in the interior and the Yukon Flats National Wildlife Refuge and throughout the southern part of the Arctic Refuge area -- that general region. And it is very real in a different way than in the lower 48. If you travel to towns and say you are interested in global warming, every person you talk to just about has something to say about it."

Gwich'in Moses Lord of Fort Yukon said, "There's a big change in the land with the melting permafrost."

Williams said, "We have collapsing highways, collapsing homes, you will see all of that when you come to Alaska. You have tipping light poles and so forth."

Charles Wohlforth said, "The softer permafrost is causing streams to change course, and causing lakes to break through into one another."

"It is quite evident that the permafrost is being affected," Climatologist Dr. Josefino Comiso said. "There are reports of electric pole lines toppling because the anchor was in the solid permafrost, but that is melting. And many parts of Alaska have been damaged because of the movement of the permafrost underneath."

Chapter 16

Vanishing Tundra

While forests in southern Alaska are dying, they are encroaching into formerly tundra and brush covered regions in northern Alaska. With increasing temperatures and precipitation, the tree line continues to move north, within a few dozen miles of the Arctic Ocean. This encroaching change means the elimination of the tundra, and questions the survival of the animals that depend upon that fertile wetland ecosystem.

The Greenpeace report "Vanishing Ice" describes the importance of the coastal plain. "Birds migrate here from six continents to nest on coastal tundra. Millions of seabirds raise their young in colonies along the water's edge. Streams and coastal spawning grounds contribute to the rich fisheries of this region. Caribou wander the tundra, also home to wolves, fox and a variety of small mammals. These living resources are likewise vital to the survival of coastal people."

A World Fund for Nature report in April 2000 warns climate change could eliminate 50 percent of rare Arctic bird populations. The report states that higher temperatures will cause wooded forests to advance northward, replacing the Arctic tundra, an essential breeding area for millions of birds. The study finds that the Arctic water birds most threatened by global warming include the critically-endangered red-breasted goose, tundra bean goose, spoonbilled sandpiper and emperor goose.

With a global temperature increase of only 1.7 degrees by 2070, all of these birds would lose more than 50 percent of their habitat, the report notes. More than two-thirds of all geese and nearly 95 percent of all calidrid waders breed in the Arctic. The study forecasts that a 40 percent to 57 percent loss of tundra in the next 100 years may decrease habitat for 5 million geese and 7.5 million calidrid waders.

"This study once more underlines the urgent need to reduce the emissions of global warming gases to slow the rate of climate change," the researchers write.

The vanishing tundra means a loss of habitat for all tundra species. As the forest marches north, this could mean a loss of open tundra habitat for migrating birds, musk ox, caribou, and more, said retired Arctic National Wildlife Refuge biologist Fran Mauer.

"Many of those (bird) species are adapted to open tundra environments. They don't nest in forests. You don't find tundra-nesting birds nesting in the forest already. So when we wonder if they would do so if there were no other place, you see the problem is that this isn't going to happen in one year's time. The forests aren't going to pop up on the North Slope in one year such that these tundra nesting birds would fly up there next year and find a forest there and have no place else to go.

"It would happen more gradually and populations would make adjustments accordingly with the gradual movement of forests northward. With tundra habitat shrinking you can only support so many birds per unit area and eventually those bird populations would probably shrink. So in the long run, some species that require tundra habitat may either have to adapt or disappear or go somewhere else. If there is still tundra somewhere else, like the Canadian Arctic islands, some of those species might persist in those areas. But the total populations would be less than they are now just because of the shrunken amount of habitat available. It is just very hard to get into specifics about this stuff because we've never lived through it before," Mauer said.

Biologists also warned of impacts to Muskox, which are remnants of the ice age with curved horns and hair reaching to the ground that is eight-times warmer than sheep's wool.

"I don't know of any significant musk ox populations living and surviving in forest country. We do see them in tree-line settings sometimes, both in northcentral Canada and we've had some musk ox come across the south side of the Brooks Range where there are sparse forest communities. But I don't know of any thriving musk ox populations in continuous forest lands. More forest would probably shrink habitat for musk ox as well. Tundra or grassland open country is what they seem to need and are evolved for. Living in forests probably wouldn't work for them."

Deborah Williams said, "With the vegetation marching north, not only does that change the vegetation, but you also have animals following that vegetation, such as moose on the North Slope where they had never been seen before.

"Moose have only reached out on the North Slope in the 20th century, since about mid-century," said scientist Dr. Glenn Juday. "In the mid-1950s people began to report them. The forest ecologist said moose are now well established with the loss of tundra habitat with the establishment of the tall shrubs that is their food source. "That is likely caused by the warming."

Our next destination was the North Slope, and I was giddy with excitement. It had been 11 years since I was there last. The place engrained so deep in my soul from my Arctic odyssey in 1991 would soon envelop me again.

Dirk flew in in his blue Beaver single-propped plane, and it wasn't much time before we were loading gear, and rolling fuel canisters over to the plane. Lifting off, we flew up the Chandalar River, snaking through the landscape. To the west, I saw Misty Mountain which I had climbed in my last journey to Arctic Village. To the east, I gazed upon Old John Lake, which I would backpack to after returning from the coastal plain.

I was in utter awe, my eyes glued to the windows as we flew through the Brooks Mountain Range, with untouched mountains, cascading rivers and sparse forest. The vegetation reached up the mountains, changing to brush, then tundra higher up the mountain, next to colorful lichens and then bare rock. Waterfalls roared out of mountain cliffs. We landed to a bumpy stop on a tundra strip in a wide glacier-carved valley, along the Canning River. We would stay here a night, then fly on to Sunset Pass.

Walking across the tundra, I noticed immediately that it was far drier than I had ever seen it in my entire summer in the Arctic in 1991, and my journey in 1993. Rather than being moist and full of life, it was dry and dead. The dry conditions prevailed because evaporation had increased with the warmer temperatures, and the permafrost had melted deeper, dropping the water table, despite the increased precipitation caused by climate change. I also noticed more brush and trees. I had photographed a small forest in the area in my 1991 trek as the tree line move north.

Camping along the Canning, I remembered my first epic journey across the Arctic, when I followed this river up to its headwaters, living on the fish that I caught. Dozens of grayling caught with hour upon hour of patient casting fed my starving body, as I carried an 80+ pound pack including a raft. Now I enjoyed the scenery, not feeling a need to fish the waters as I had brought enough food. Just being in this place was so fulfilling, but I bristled at the threat looming just across the river.

The Canning forms the western boundary of the Arctic National Wildlife Refuge. Camped on the east side, we were in protected territory. But on the other side, just downriver was state-owned land leased to oil companies. Farther west is the massive, sprawling oil development of Prudhoe Bay.

The next morning, we flew on to Sunset pass. Flying through the mountains was like living in a dream: I stared out the window, constantly looking around to try to avoid missing something. Mountains, waterfalls, glaciers and rock monuments adorned the landscape in all directions. One mountain glistened in the sun, completely covered with snow and ice: reflecting a white brilliance of legendary beauty.

We flew over Lake Peters, which had glacier silt mixed in with its blue waters from the Chamberlin Glacier perched on Mount Chamberlin just to the east. We went over the channel to Lake Schrader, which I had rafted in 1991, subsisting on ten-pound lake trout that I caught aplenty. Schrader was a clear blue color, with a streak of glacier silt coming downstream from Lake Peters.

Seeing the land from the air, the whole of the place becomes visible. In a much greater way, scientists now have satellites to look at the entire globe, so the reality of climate change simply cannot be denied. It has been measured across every square miles of the globe for decades, and scientists warn that the changes are unprecedented.

Chapter 17

The Big Picture

Scientists can say with much greater certainty that climate change is real in Alaska in large part because of satellites. The irony of satellites is that a technology that uses an enormous amount of energy to get into orbit is alerting us to the need to conserve energy and switch to renewable sources.

“Climatologists have been hindered until the advent of satellite data, where you really have the global coverage of data every day,” said Climatologist Dr. Josefino Comiso. “So they are making breakthroughs now all of a sudden with this satellite data. As we get wiser and wiser in introducing some of the complications then we come up with better and better models, and get to the point where we are able to predict what will happen with the Earth’s systems.”

Comiso also confirmed major increases in temperature in Alaska. “We observe that over the entire Arctic region, North America shows the highest rate of warming, and that includes Alaska. What I figured out as I looked at long-term data sets, looking at surface data or meteorological data from 1900 to the present, the current trend toward warming as we observe from satellite data is about eight-times higher than that of the 100-year data set.” That is an unprecedented upswing in the rate of temperature change.

"I focus my studies mainly on polar regions. I do polar regions because that is where satellite data can really help. There are only about 20 meteorological stations in the polar region, whereas if you look at England or Germany, or the lower 48 states of the US, you have meteorological stations all over. They have a very good idea already that there is significant warming. I look at satellite data and this really proves that the global climate is warming.

"We are consistent. It is only with satellite data that we can have a true global picture of the climate. Looking at Alaska, for example, some areas might even be cooling compared to other areas, such as the Bering Sea region. The sea ice has been expanding in the Bering Strait in the winter. Major meteorological systems in Alaska may be different, and if you try to make deductions from one area, you miss the greater picture.

"With all of these complications, even with the greenhouse effect, it will not be uniform across the Earth, it may be more in some areas than in others. And therefore, the heating of the Earth can be a very complex thing. In Alaska, there is an earlier spring and a later freeze-up. That would be consistent with permafrost melting more and more."

Climatologist Uma Bhatt said that while the Earth's climate varies in a number of different cycles, some within a decade, some multi-decadal, that the input of greenhouse gas emissions clearly has contributed to a sharp increase in temperatures recently.

"What is happening is the storm tracks, where the main precipitation band that goes across the North Pacific and comes into Alaska, ends up getting shifted northward. There is actually less precipitation in the south part of Alaska. The main precipitation in the band that heads across the Pacific and hits mountains in Anchorage and Juneau is reduced because now there is higher precipitation further to the north."

"If the precipitation moves it does not necessarily mean that Alaska will become greener," Bhatt continued. "So it is very complicated. It would probably be drier in the southern part of Alaska and wetter in the northern part." Because the climate models aren't fine tuned, she said, it isn't possible to predict precisely how the precipitation will change.

Despite the critical need for more weather satellites to study the changing climate, NASA has cancelled or delayed several critical satellites designed to study global warming, including the Deep Space Climate Observatory (cancelled) and the Global Precipitation Measurement satellite (delayed). Instead, NASA has prioritized sending people back to the moon by 2020.

From the amazing view in the bush plane, I followed the Kikiktat River from Lake Schrader down toward the Arctic Ocean. This was the river I nearly died in after a rafting disaster in 1991. We flew over Sunset Pass, then landed on a flat gravel area by a small stream. We moved our gear from the plane.

The pilot lifted off from the gravel runway to fetch the second group from our previous camp, a large male caribou came running up with a huge rack. We all snapped photos and kept a low profile. The caribou did not seem to notice until about a 100 feet off, as it was headed downwind, coming from the coastal plain to the North.

Two months prior, tens of thousands of caribou had massed together on the plain, giving birth to 40,000-50,000 calves. Yet this is where some still want to drill for oil, in this last corner of wild America at the shore of the Arctic Ocean: the last frontier. It is emblematic of our place in history: the end of the fossil fuel era. We have already taken all of the easy to get oil, and are now subject to shortages that can spike prices to levels not imagined before.

More importantly is what this means for our climate. All of that carbon that had been stored over hundreds of millions of years, that acted to cool the planet to the point where intelligent life could evolve, was now being pumped into the atmosphere, causing an irreversible shift back to the hellish conditions of tens of millions of years ago that our biodiversity today is not evolved to survive. A 2006 study in the journal Nature found that year-round average temperatures were 74 degrees near the North Pole 55 million years ago.

I joined a small group that hiked up a mountainside overlooking the coastal plain. Scientists report that the increase in freeze-thaw cycles—the number of times that the ground freezes then thaws out each year—will increase avalanches in northern mountainous regions like Alaska.

From the north-facing mountainside, we could see all the way out to the Arctic Ocean over a glorious wilderness in need of permanent protection. Trying to capture the all-encompassing scene by camera, I was struck by how many photos it would take to show the full picture I experienced. Each photo is but a tiny fraction of the whole scene. To have an all-natural landscape with pure, untouched mountains and undamaged streams and rivers from the headwaters where I now sat all the way to the Ocean is a jewel we must protect. With 123,000 caribou that mass together where I gazed, that feed the Gwich'in people whom I had just visited, this is a place that must stay wild.

Chapter 18

The Arctic Refuge: The Back Side of the Moon or the Last Great Refuge?

Sitting atop the northern face of the Brooks Mountain Range near Sunset Pass, I overlooked the coastal plain of the Arctic National Wildlife Refuge, which stretched far to the east and west, and north to the dark-blue waters of the Arctic Ocean. To have an all-natural landscape with pure, untouched rivers and streams from the headwaters high in the mountains, all the way to the ocean is one of the most important treasures we have. But it is so easy to lose.

Caribou were migrating from the plain back toward their wintering grounds over the mountains. One ran right up to us, a husky male with a giant rack. Seeing us, he snorted and bolted back, then ran far around. The importance of this place cannot fully be told in words. There is just a deep, unquestionable perfection in the complete, working ecosystem, and the words and charisma of the Gwich'in peoples living as part of it.

The Arctic National Wildlife Refuge contains the last five percent of protected North Slope of Alaska: the coastal plain. From a range the size of California, caribou migrate across the Brooks Mountain Range to mass together on the coastal plain and give birth to the next generation of caribou.

More than a thousand U.S. and Canadian scientists called on former President George Bush to protect the Arctic National

Wildlife Refuge from oil drilling on February 14, 2005. In a letter to the President, the scientists questioned assertions that oil could be safely extracted from the Refuge, and urged President Bush to "support permanent protection of the coastal plain's significant wildlife and wilderness values."

Scientists found that oil development could seriously harm caribou, polar bears, muskoxen and snow geese—among other wildlife. They warned it could disrupt the fragile ecosystem of the coastal plain, which could lead to more widespread injury to wildlife. The signers categorically rejected the notion that the impacts of drilling could be confined to a limited footprint, as pro-drilling forces claim, noting that the effects of oil wells, pipelines, roads, airports, housing facilities, processing plants, gravel mines, air pollution, industrial noise, seismic exploration and exploratory drilling would radiate across the entire coastal plain of the Arctic Refuge.

Defenders of Wildlife President Rodger Schlickeisen said, "hundreds of scientists are telling President Bush that throwing the Arctic National Wildlife Refuge open to oil companies will harm wildlife and permanently disrupt the wild nature of this unique place. It simply does not make sense to destroy the Arctic Refuge for oil that won't lower prices and won't make a noticeable dent in our dependency on foreign energy, when it's so much easier to get the same amount of energy through common-sense conservation steps."

Increasing the Corporate Average Fuel Economy (CAFE) standards to 39 miles per gallon, would save more than 100 times the oil than could ever come from the Arctic National Wildlife Refuge. In fact, more oil would be saved in increasing the CAFE standards than is in the Arctic National Wildlife Refuge before even a drop of oil from the refuge would get to market. With CAFE standards now set to be 35 mpg by 2020, we clearly don't need Arctic Refuge oil. These standards still need raised to combat climate change.

We have already spent about a billion dollars on the program for a new generation of vehicles in the 1990s to develop super-efficient cars, like the Precept that GM produced in 2000 getting 80 miles per gallon. Each of the big three auto companies had such vehicles. Had they chosen to produce these efficient vehicles, rather than SUVs and Hummers, they would not be in the bankrupt state they are today. Hopefully they will learn their lesson, and start producing electric mini-cars with solar panels on their roofs, as well as 70+ mile-per-gallon cars. They also should produce mass transit vehicles.

The change in efficiency standards will create far more jobs in retooling the auto industry compared to the few automated oil industry jobs if the refuge were to be opened. But it will have zero impact on people's lifestyles, except to save them money on gas and clean up the air they breathe. This could easily be done without impacting peoples choice of vehicle. Mid-sized cars can easily get 50-90 mpg.

Meanwhile, the impact of drilling on the lifestyles of those living in the Arctic National Refuge would be huge. These caribou feed 17 Gwich'in villages. The Gwich'in are caribou people, who have lived for 30,000 years off the Porcupine Caribou Herd that calves on the coastal plain, right where they want to drill for oil. These are among the last native peoples on Earth whose culture we have not yet destroyed.

This is the densest area for polar bear dens in the world, home to one of the last refuges of muskox, grizzly bear, Arctic wolves, Arctic fox, and much more. This is not just some place way up north, it affects everyone in every state of the U.S. and 6 continents around the world. More than 160 species of birds from 6 continents and all 50 states breed on the coastal plain of the Arctic National Wildlife Refuge.

But the refuge is still threatened. If drilling were to be approved, and we lose in the court battles that would ensue, this last refuge would be gone forever. To compare the Arctic National Wildlife Refuge to the backside of the moon, as pro-drilling Senator Larry Craig, a Republican from Idaho did in March, 2005, is like comparing life and death. The Arctic Refuge is among the last places on Earth with all of its pieces still intact. All of the plants and animals that were present before the human-caused mass extinction event we are currently in, are still there, making it among the most important areas in the world to protect.

While politicians have claimed drilling in the Arctic Refuge could somehow make us safer from terrorism, by providing more domestic oil, it in fact would increase the risk of terrorism, as it would put more of our energy supply through 700 miles of extreme vulnerability: The Trans Alaskan Pipeline.

A drunk hunter shot a hole through the pipeline in October, 2001, spilling 285,000 gallons of oil. Crude oil gushed out of the pipe at a hundred gallons a minute, spraying for hundreds of feet. The one bullet shut down the pipeline for a week. That is not a secure way to get our energy. Solar and wind, however, are decentralized, and very secure from terrorism.

Chapter 19

Opening the Floodgates: The Arctic is Just The Beginning

In 2005, Conoco-Phillips joined BP in pulling out of Arctic Power, the lead lobbying group working to destroy the refuge. The benefits to oil companies for passage of drilling in the refuge extend well beyond any windfall in the Arctic. As one proponent of drilling said, opening the refuge would "break the back of the environmental movement," and open up the coasts of Florida and California to drilling, where there is a significant amount of oil.

On March 17, 2005, Senator Ken Salazar, a Democrat from Colorado, said, "last night on the Senate floor, I continued my determined fight to say to our children and grandchildren that we will protect certain pristine areas of this great country. The Arctic National Wildlife Refuge is one of those places." He said that the Arctic National Wildlife Refuge is "a refuge from encroachment and a symbol of restraint and our optimism for the future. Today I voted for an amendment to protect the refuge, and yet it is with a heavy heart that I write this statement with one of our last remaining wilderness areas under siege, threatened to be opened up for exploration and drilling for a six-month supply of oil. This move would set an ominous precedent."

Senator Salazar continued, "The Coloradans and Americans alike should know that once this door is unlocked it will be that much easier to unhinge the protections of every refuge from the

Alamosa National Wildlife Refuge in my home San Luis Valley to the Florida Everglades Wildlife Refuge. From the purple mountains majesty to the coasts of our shining seas. It is my hope that we can demonstrate to the American public our commitment to future generations through protecting pristine lands. We owe it to our children as a compact to the generations beyond."

There is a pervasive picture of Americans as gas guzzlers who just don't care about the environment, and that it is the general public that is pressuring politicians to get oil cheaply, regardless of environmental costs. It's a convenient scapegoat: blame the general public. But on this issue, most Americans see the facts clearly—and they don't like what they see.

A bipartisan national survey found that by a margin of 58 to 33 percent, Americans oppose proposals to drill for oil in the Arctic National Wildlife Refuge, according to U.S. Newswire, February 9, 2005. The bipartisan telephone poll of 1,003 registered voters was conducted January 13-17, 2005. Almost twice as many people opposed drilling than favored it.

A Democratic and a Republican polling firm teamed up to do the national poll. Democratic polling firm Lake, Snell, Perry and Associates Spokesperson Celinda Lake said, "Voters believe there are some places that should simply be off limits to oil drilling and the Arctic Refuge is one of them. They believe we have a moral responsibility to protect this unique area, and the abundant birds and wildlife that live there for future generations."

The spokesperson for the Republican polling firm, Bellwether Research and Consulting, Christine Matthews said "one of the most striking findings from this poll is the degree to which voter opinion on this issue of drilling in the Arctic Refuge has solidified, moving from the realm of public policy issue to value. Only about ten percent of Americans are undecided on this issue—most people know where they stand when it comes to drilling in the Arctic National Wildlife Refuge.

Even more than the number who opposed drilling, a resounding 73 percent of Americans said that the issue of drilling in the Arctic Refuge is too important to the American public and future generations to be snuck through in the budget process, and disagree with drilling proponents' arguments in favor of allowing drilling to be determined in the budget. Even among Republicans and those who voted for George Bush in 2004, a majority oppose using the budget process to open the Arctic Refuge to oil drilling. The opposition is consistent across age, gender, party and other demographic groups. Dozens of other national polls over the last two decades have found a vast majority support refuge protection.

"Ultimately, the battle over the Arctic Refuge says a lot about what sort of nation we are going to be," read a statement released November 14, 2004 by fourteen environmental and Native American organizations. "Are we a country that squanders its natural resources for short-term profit? Or are we a nation that stewards those resources, carefully conserving the most valuable natural and economic assets to ensure we will have them in the future?

"For some five decades, thousands of citizens from all walks of life and all areas of the nation have worked to protect the Arctic Refuge. Leaders of both political parties have recognized the importance of protecting the area. President Dwight Eisenhower established the Arctic National Wildlife Range in 1960. It was enlarged by President Jimmy Carter in 1980. Backed by the majority of American people, bipartisan majorities in the US Senate have voted time and again to keep the oil drills out of the Refuge's Coastal Plain.

"Protecting the Arctic Refuge is important ecologically and culturally. The narrow Coastal Plain is the foundation of a delicate ecosystem that features a spectacular diversity of wildlife, including 160 bird species, polar bears, musk oxen and wolves. The area also supports the Porcupine Caribou Herd, which calves on the Coastal Plain. The caribou are the basis of the subsistence culture of the Gwich'in people, whose communities lie along the migration route of the herd."

Senator John Kerry said, "The fact is the price of oil will not drop, the price of energy will not drop, the price of gasoline will not drop, because with three percent of the oil reserves of the world in our hands including Alaska, you can't drill your way out of America's predicament."

The key is in conservation and investing in renewable energy sources like solar and wind. With just a fraction of our military budget invested over a few decades, we could be entirely powered by clean solar and wind energy. As we invest more and more in solar and wind, we can replace the inevitably declining supplies of fossil fuels, that we really should not be burning anyway.

Our sunset pass camp was quite peaceful, with the gurgling sound of a clear creek nearby. I guided the group of students on a hike toward the Saddlerochit. Harvard and the two outdoor store owners, Haines Boren and John Cooper accompanied us on the beginning of the hike, through the wide u-shaped valley of sunset pass. We followed the clear creek, going up and down ledges that the creek had made ages ago. Dense willows made crossing small streams challenging. Ten caribou came by, followed by a bull and cow on the horizon, with the silhouette of Mount Chamberlin in the background. A small, young caribou walked along the crest of a mountain. We encountered dozens more caribou.

The same willow bushes that at times made for difficult hiking also threatens the future of the caribou. Brush and trees are growing larger and in more areas in the changing climate, reducing the tundra habitat.

Chapter 20

Locking Horns with Climate Change: The Caribou Challenge

Fran Mauer warned that a forested North Slope, as Glenn Juday predicts if current trends continue, would be a drastic change for all tundra wildlife, in many cases threatening their continued existence in those areas. “It would be really different than it is today and has been over the time period that we are familiar with. It is hard to fathom the whole thing. What would it be like if there was a forest on the North Slope? The relationship between predators and prey would be vastly different.”

He said that with northern migratory caribou herds, “a big part of their behavior and their migrations is based on the open terrain that they live in and their ability to see and detect predators from a distance. All that strategy would not work well in a forest situation.

“We do have caribou in North America that live in forests and we call them woodland caribou and they behave much differently. They don’t travel in big massive herds. They behave more like white tailed deer on a spectrum of behavioral characteristics. The caribou phenomenon as we know it today would be very different if it were mostly forested. I can’t see the forests going high in the Brooks Range to the tops of the mountains in the near future by any means, so there would be some higher elevation tundra areas more fragmented and separated by forested valleys so conceivably there may be smaller less migratory isolated subpopulations of caribou scattered in the Brooks Range.

"That is similar to what we see in southern Alaska, where we don't have any large migratory herds in southern Alaska: they are relatively small, isolated herds. And many of those herds are declining. Some of them have declined dramatically and are endangered."

Mauer said that caribou have been acting differently since the late 1980s, spending less time on the coastal plain and in two remarkable years they calved prior to reaching the coastal plain.

"Why the calving didn't occur in the Arctic coastal plain in 2000 and 2001," Mauer said, "may or may not be related to climate change or global warming. What happened in both years was there was a very late onset of spring, and a fairly deep snow on the winter range and along the migration route northward. The migration of pregnant females was delayed by as much as a month. The deep snow further slowed them down in moving north. They tend to not move until there is definite melting underway. If there is cold inclement weather during migration, a lot of them stop moving.

"That is the kind of conditions that occurred in 2000 and 2001. We had the extreme case of caribou not making it onto the Arctic coastal plain coming out of the Canadian winter ranges primarily and having calving scattered along the migration route and off in the Canadian region. We saw an extreme in that pattern. We've seen that kind of thing in other years where the spring is late, but not as pronounced as what we saw in 2000 and 2001 where there were even a few cows that were born south of the Porcupine River which we had never seen before in 30 years of monitoring caribou there."

Mauer said there was "extreme late snowmelt. I wouldn't say necessarily record deep snow but deep snow on the migration route. Now whether that is due to some global warming connection I have no idea. Some people would say well that certainly is no sign of warming, we have later springs and slow melting conditions. But this whole business is very complicated, and we may see extremely different patterns than we are used to but not always warmer conditions."

"It is different, and how well they are able to adjust to it, I don't think anyone has a solid answer to that," Mauer said. "It certainly is different. For whatever it's worth, the Porcupine Caribou herd has been in a declining trend. That trend began at the end of the 1980s, which was about the same time that we began to see this different movement pattern on the coastal plain. Whether they are related or not, I have no idea. There are many things that can influence the population level.

Alaska Coalition Outreach Coordinator Dan Ritzman said that the increase in mortality during those years accentuates the importance of protecting the Arctic Refuge coastal plain from oil development. Should development be permitted, caribou would likely avoid those areas as they have around Prudhoe Bay. This would push them to calve into the areas that caused such a loss in population.

Mauer concurred that there are similarities between how oil development would impact the caribou. “Yes they did have higher mortality of calves in those years. If there is oil field development, the caribou will get there, but once they get there they will shy away from the development zone and have to seek alternate places, often going near to the mountains.

“The problem is the oil field development would be there year after year after year, and it would be a permanent loss of calving area. Whereas the extreme weather conditions of 2000 and 2001 have swung back in the last couple years. It was an extreme example of what we could expect, but it is not exactly the same.

“In other years when caribou are late, the whole herd still moves onto the coastal plain. Another difference with the weather phenomenon in 2000 and 2001 is it is a good thing they did not make it to the coastal plain because it was covered in snow. The caribou adjusted to avoid going out on the snowy coastal plain. They at least gave birth to their calves where the snow was melting.”

If the caribou are squeezed toward the mountains by oil development, this would cause an increased predation on calves and a decrease in food supply, Mauer said, both of which would negatively affect the Porcupine Caribou Herd. “There is a greater risk of predation. Wolves have their dens on the northern mountain valleys, not on the coastal plain. Most of the bears den in the mountain foothills, not on the plain. Up near the mountains there are more golden eagle nests. When you add all those up, it puts the calves in closer proximity to predators. So we expect there would be a greater loss of young calves when they are the most vulnerable. In studies in the early 80s, we found a great risk to calves that spent time near the mountains.

"Also, if they are squeezed to the mountains, there is more of a limitation of food. Their options of food are more constrained in that they can't go out in the plain following the plant chronology. That will likely result in a slower growth rate of calves and influence the health of the calves."

"The last time the count was done was 2001 and there were 123,000 caribou counted on aerial photos. The peak count was in 1989 when 187,000 were counted. So it is a drop of more than 60,000 in twelve years.

Satellite data found that vegetation has increased by 10 percent over the last 20 years in the Arctic. This increased brush makes it harder for caribou to move through their grazing grounds. Caribou are the farthest land migratory animals on earth, traveling 2,700 miles per year from their wintering grounds to their calving grounds.

We experienced the challenge of having to forcibly push through dense willow brush on our hike. This added obstruction is a major issue for the caribou. Also, caribou must worry about lurking wolves or bears in the brush which they otherwise could see from a distance in the open tundra, as well as the increased effort to get through it.

Mauer said that though the caribou may stay on the coastal plain for a shorter period of time, that does not diminish the need to protect the plain from oil development. "As long as caribou do go on the coastal plain, we ought to remember why they are going there. The main thing they do is give birth to their calves. From a vulnerability standpoint to the calves, this is the most critical time. It is when calves are incapable of defending themselves from predators. As long as caribou are going there to give birth to their calves, it is going to remain a serious issue if it is open to drilling."

Kaktovik resident Robert Thompson reported in 2004 that hundreds of caribou died on Barter Island after a rain storm in the winter created a layer of ice on the tundra. Caribou normally kick through the powdery snow to feed on the frozen tundra. But the layer of ice made it impossible for them to feed. While historically rain was unheard of in the winter in the Arctic, with climate change such events are now common.

It is symbolic and ironic that the caribou are being attacked on both fronts in this last frontier in the last corner of America. Climate change and oil development are related and are both threatening the survival of the greatest nomadic land animal on earth, along with the Gwich'in Nation that depends upon them.

Hiking through the marshy terrain was challenging, but it was much drier than I had encountered during my first expedition. The group of students had read my first book, Arctic Quest. Several were interested in catching and eating some fish, as was I. But I also had some reservations, and a strong desire to just leave this place be, having taken so many fish to survive during my first journey, and now having plenty of food for my short stay.

We got to the edge of the Saddlerochit, River and I could see where the Kikiktat River entered into it. I followed the Kikiktat upstream with my binoculars until it disappeared into the mountains, knowing Lake Schrader was not far away. I was near where I nearly perished during my first journey, when my raft overturned on the flooded Kikiktat, and I was without a life preserver or a wetsuit.

I found what appeared to be a good fishing spot, but not a good camping area. We settled for wet, tent locations for a good chance at fish. Two of the students, Nathan and Landon caught seven grayling between the two of them. I filleted them, and Landon cooked them. Eating a delectable meal of fresh fish with seasoned mashed potatoes, we enjoyed an amazing view of Mount Chamberlin, with the red glow of the semi-setting mid-August sun.

Like nearly everything in this far northern land, the grayling too are threatened by the changing climate. As the permafrost melts, more nutrients enter the Arctic waterways, making them more like temperate waters, he said.

"There is some research going on at Atuvik Lake field camp about what happens when you greatly increase the nutrients in the stream," Wohlforth said. "Because when you have the thawing permafrost going into the stream, you have a lot of nutrients being released into that water… A lot of research shows that the biological ranges are moving northward." The big question is what will happen to the species that live in the Far North: the Arctic Char, Dolly Varden, Grayling, Burbout and others as their Arctic habitat changes more temperate.

Polar fish species are particularly sensitive to global warming, concluded a study published in Science, January 5, 2007. Warming causes a deficiency in oxygen uptake and supply in Arctic fish that limits their growth and reproduction. The study found that fish in Polar regions have an especially small thermal tolerance window.

Another major impact from climate change is that snow packs and glaciers that once fed streams all summer long are melting away, drying up the creeks and killing the fish.

During my 1991 expedition, I wore a thermometer on my jacket. I experienced several snowy days through August, and the temperature did not exceed 70 degrees Fahrenheit. During this 2004 journey, we experienced temperatures in the high 80s. In 2005 I also experienced temperatures near 90 degrees.

The Arctic Climate Impact Assessment Team, a group of 300 scientists and indigenous peoples, found in 2004 that year-round average temperatures have risen by five degrees Fahrenheit in Alaska since the 1960s, and average winter temperatures soared eight degrees in that period. Scientists at the Intergovernmental Panel on Climate Change predict the Earth is expected to warm up to 6.4 degrees Celsius (11.5 degrees Fahrenheit) by 2100.

"During the last decade, record high annual average temperatures were broken at 14 of 38 reporting stations, while record low annual average temperatures were not broken at any of the stations. No record monthly or annual low average temperatures have occurred at weather stations in Alaska for decades" (Knight Ridder, 2003).

The western Arctic is one of the fastest warming regions in the world, warming at a rate of one and a half degrees per decade for the past three decades, several times the global average. Temperatures are up to 25 degrees higher than normal, according to the Christian Science Monitor.

New York Times Reporter Timothy Egan reported that the average temperature in Alaska has risen by seven degrees in the last 30 years. Climate models predict that temperatures may rise by as much as 18 degrees more this century.

The Arctic Climate Impact Assessment Council Report released in November, 2004 found that the Arctic is particularly sensitive to warming because its plants and soil hold less water than more temperate environments. Thus, more of the sun's energy is used for heating the surface rather than evaporating water. Other temperature zones shed heat by shipping it north in ocean currents and wind patterns, but the Arctic is the end of the line on the planet.

Chapter 21

Witnessing the end of a glacier

In 2005, I also guided a group of Appalachian State College students through the Arctic Refuge. This time, I brought a video camera. Like the previous year, but with a different group of students, we visited Old Crow, then flew into the refuge, again going from Sunset Pass to the Saddlerochit River. This time, we backpacked around the north side of the mountain that lay in between, climbing high up onto the loose rock, which was still easier than the marshy tundra.

Again, temperatures were in the high 80s: unbearable to the Arctic natives. Camping on the banks of the Saddlerochit River, the group caught many grayling. I focused on getting footage with my video camera. A grizzly bear wandered along a creek, headed our way. The bear ducked in and out of the willows that grew along the stream banks. The grizzly moved fast, and we were all apprehensive. Being with a group of seven, compared to my spending months alone in this wilderness during my 1991 expedition without even pepper spray, I reassured myself that all would be well, as it was.

But as we lay in our sleeping bags, with just a thin piece of fabric between us and the outdoors, our vulnerability to the elements – and the bears – was accentuated.

Though we may live in more secure homes than tents, we are still vulnerable to the climate. We depend upon it to bring us drinking water and to nourish the crops we need for food. The delicate balance of our climate is critical to our survival in a myriad of other ways, such as the plants that provide the oxygen in the air we breathe.

The morning of our last day, I lit the stove to cook the group's fish. Shortly after lighting it, the stove sputtered out before the fish were done. I wanted to make a fire, but little wood was nearby.

This is what is going to happen in coming years. We will run out of easy to get oil and natural gas, right when we need it most to finish the transition to all renewable power. It will sputter out of the ability of most Americans to afford.

I suggested that we pack up the fish and stop to cook them on the way when we got to a place with sufficient wood for a fire. After a couple of hours of backpacking, we made a fire near a creek, with everyone pitching in to get wood. Soon we were enjoying the fish by a fire and a cascading, crystal clear creek, as we rested from backpacking. What seemed a bad thing of running out of gas turned into a nicer setting and tastier fish cooked over a willow fire.

Several people had blisters, including myself, and I showed the group how to chew on willow bark to reduce the pain. Willows have the same active ingredient as aspirin: salicin. "Not the tastiest, but it does work," I said. In my previous journey, I cringed at the damage that I was causing, knowing it can take centuries for willows to get a few inches thick. But with climate change massively increasing willow growth, to the point where it now hinders Arctic animals, I considered it miniscule mitigation.

Shortly after we ate, we climbed up a narrow valley toward a pass between two mountains to get back to our camp. We came upon the last remnants of a glacier, about the size of a van. I took video, and photos, and encouraged the students to do the same with their cameras.

"This is a tragic sight, and one we must bring to the masses," I said. "This is the last remnant of a glacier that once filled that vast area," I said, motioning to the barren moraine that once was covered by glacier.

We climbed over the loose rock that had been covered by ice, to reach the summit of the last ridge before the valley of our base camp. The glacier had been more than a mile across.

I still hear people say that climate change is just natural. They point to the ice ages, and say look, we warmed from that, of course there is climate change. But the phenomenon referred to as climate change today is human caused. We are contributing to the vast majority of the change, unleashing a potentially unstoppable switch into a hellish climate of death and destruction. Those depending upon the water that the melting glaciers once generated will now go without. With all of our technological might, we still need clean water and air to survive.

Chapter 22

Human-Caused or Natural?

While many people agree that there are major changes in the weather, not everyone understands the solid scientific proof that the burning of fossil fuels is the primary cause of the climate crisis.

Climatologist Uma Bhatt said, "The other night I was sitting at an American Meteorological Society meeting in Seattle with a bunch of scientists, and one of them said 'you know 15 years ago if you asked the average scientist if you see a human impact, 50 percent of the people would have said yes.' But now if you asked I would say most atmospheric scientists, probably 95 percent would say yes, because in the last 15 years there have been a lot of pieces of evidence that have come in. It is not just looking at one time sequence, it is looking at all of the things that come in. And for me one of the most compelling pieces of evidence is if you look at nighttime temperatures. Because then that is a radiational impact. You see a strong significant trend in nighttime temperatures globally. That is where you really see it. So I would be very comfortable in saying that we do see human impacts on the climate."

Climatologist Dr. Josefino Comiso said scientists are convinced that human greenhouse gas pollution is to blame for the unusual weather. "We are now in a position I think where we can construct models a lot more accurately than 10-20 years ago. We have a much better understanding of physics than before. We have much better satellite data than before, not just in resolution but in quality. We have much more comprehensive data sets. The data that climatologists could only dream of 20 years ago is now available.

"We can say with a lot more confidence the impact of carbon dioxide and other greenhouse gases. It's indeed a big impact. Even if you consider the large variability and the Arctic Oscillations, we show a much greater association of greenhouse gases with the warming."

Guntner Weller said, "I am also totally convinced that anthropogenic is a significant (cause of) climate change, and this is the consensus of most scientists from around the world."

In reply to those who deny climate change or its negative affects, Patricia Cochran, executive director of the Alaska Native Science Commission said, "They obviously don't live in the Arctic."

Former Boston Globe Reporter and Editor Ross Gelbspan wrote a book in 1997, "The Heat is On" about the fossil fuel industry efforts to discredit climate change. He found that the conclusion that human-caused greenhouse gas emissions are causing the planet to warm up are the result of the "most rigorously peer-reviewed scientific collaboration in history. The contradictory statements of a tiny handful of discredited scientists, funded by big coal and big oil, represent a deliberate—and extremely reckless—campaign of deception and disinformation."

If only more leaders would have heeded the warnings earlier. Now we must hold those who spouted the misinformation accountable for the mass murder they have caused by the delays.

On March 1, 1999, the American Geophysical Union issued a strong position statement on climate change. The group, representing 35,000 Earth and planetary scientists, warned "there is no known geologic precedent for the transfer of carbon from the Earth's crust to the atmosphere" in the amounts comparable to fossil-fuel burning without simultaneous changes to the climate.

The atmospheric concentration of carbon dioxide has risen from 280 parts per million in 1800 to 385 parts per million in 2008 due primarily to the combustion of fossil fuels and the destruction of forests. "What's happening to our environment is not natural. It's a problem of our own making," said Eileen Claussen, President of the Pew Center on Global Climate Change in a November, 2004 report. "The longer we delay in reducing greenhouse gas emissions, the greater the problem will become."

"Historically, the Earth has never been over 290 parts per million carbon dioxide," said Ohio State University Geological Science Professor Dr. Lonnie Thompson on National Public Radio, December 17, 2004. "We are engaging in a giant experiment, with disastrous and often unforeseen consequences. Humanity is adding 2.6 parts per million more carbon into the atmosphere every year."

Comiso said, "the question is to what extent the increases in greenhouse gases would cause a warming. And this has been attempted by several scientists around the world who have published their results. They have reached the stage that now we are getting global data sets. We recently launched satellites that can measure parameters with greater precision.

"To understand the climate system, you have to have input numbers from different sources. But now, with the use of satellite data, we have greater interpretation of greater atmospheric parameters. (We have) greater global data set, and more powerful computers that allow us to run the model longer. Taking 8,000 years of climate data … used to take several weeks. It now takes several days.

"As we are getting able to get better and better models that are more able to represent the climate, then we are able to have better and better confidence about the real impact of fossil fuel burning on the climate. And then we are able to sort out what impact CO2 and other sources of greenhouse gases are contributing to the warming that we are seeing right now.

"You have a lot of research going on in addition to the research that we have been doing around the Earth that looks at the confidence level for all of these studies. And it is hard to establish that we are one hundred percent confident mainly because we just do not have enough data. But we are moving very rapidly in that direction. It is very clear that the weight of data shows that we are on the right track. We know that the science is there.

"I don't think that there is any question right now that the Arctic is warming. Even people who are very much against the Kyoto Protocol for example accept the fact that global warming is occurring. Unfortunately, it is much more difficult to establish that global warming is one hundred percent caused by the burning of fossil fuels.

"But we have enough data to show the probabilities are quite high that CO2 is the cause of much of the global warming. Recent modeling studies support the fact that the warming is caused by CO2 and other greenhouse gas emissions."

While some may point to isolated areas of cooling, Comiso said satellite data can sort that out and show that on the whole, the trend is toward warming. "The global climate system is very complex. With global satellite data you see warming in some areas, and cooling in others. That is just the way the system works."

But the coal and oil industries have unleashed a massive publicity campaign to discredit the reality of climate change. Fran Mauer said "a lot of the corporate interests that have a lot at stake have been actively involved in promoting confusion. The best analogy that sheds light on that phenomenon is to look at the tobacco industries' response to the growing number of studies showing the risks of smoking, and the campaigns that the tobacco industry engaged in." He said the fossil fuel and auto industries are engaging in similar tactics with climate change.

Arctic Screaming

Even the most ardent of climate skeptics, like Exxon Mobil, are finally coming around to admitting that climate change is caused by humans, and is a serious problem. But the decades of delay these industries have caused needs to be redressed.

I was the last to leave the remnant glacier. It was nearing midnight, and the group was afraid that their professor, Harvard Ayers would be worried. We still had a ways to go. Nonetheless, I stopped to get video of the area where the glacier had been.

Back at camp, I lay down, exhausted. I was carrying about 70 pounds with all of my video equipment. Though far less than my first expedition, I was not as well conditioned as I was then.

A dense fog hung over the valley, and far out onto the coastal plain. Our bush pilot could not safely land. Harvard did not have a satellite phone. I had gotten reception at one point with my cell phone up on a mountain, coming across the coastal plain either from Prudhoe Bay or Kaktovik, and made a call. Concerned about the group, and their connecting flights, I climbed that mountain every day to try to make calls. Though my phone showed a bar of service, call after call I tried, but none went through.

After nearly a week of waiting with diminishing food, the group was apprehensive. Many were miserable. I felt bad for them, but was quite happy myself to have more time in the wilderness. I walked miles every day to climb mountains over-looking the foggy coastal plain. I called, trying to reach our bush pilot.

Finally, the fog cleared and our pilot flew in and landed near camp, where we had a group waiting with their gear. I took the last flight out. We flew across the coastal plain seeing several herds of caribou, as well as some musk ox, a grizzly and flocks of migrating birds. How critical it is to leave a place as near natural as we can, without the scars of oil development. In 2005, with Bush having been "re-elected," and gas prices on the rise, drilling seemed inevitable. A deep sadness hung over my soul that came from my interviews and discussions with the Gwich'in earlier in the journey.

Like 2004, I had spent a week in Old Crow, this time with my video camera. Though they generally shunned cameras, the Gwich'in welcomed mine, knowing I would present their pleas for survival: to stop drilling in the Arctic Refuge so the caribou could continue to feed their people sustainably. The innate sadness ingrained in their voice touched deep in my soul, as I too shared a strong passion for protecting the coastal plain.

We followed the string of tiny islands to the landing strip on Barter Island, a low barrier island. As we got out, the police officer for the village approached the pilot. "How is the weather?" he asked. "The Senator is flying in."

"It is a really low ceiling," the pilot said, referring to the clouds and the amount of space under them for the pilot to fly and see the ground and runway.

"Which Senator?" I asked.

"Stevens," the young officer answered. "He is having a public meeting."

Chapter 23

My First Encounter With Ted Stevens

I rushed to Waldo Arms "hotel" to get cleaned up as fast as I could, then went over to the meeting at the community building.

Stevens held the meeting to try to show support for oil drilling, but he found quite the opposite. I entered the meeting, and Senator Stevens was talking to citizens.

I shook his hand and said I had just returned from the refuge two hours ago, and that the area needed to be protected from oil development.

Stevens said, "son, I've been working to open up that area since before you were born." In fact, Stevens spoke for the protection of the coastal plain area when he was a state senator in the 1950s. It was only when he began to accept large contributions from the oil industry that he changed his stand on drilling in the area.

Secret Service agents then tried to push me outside. I said no, this was a public meeting, and that I would stay. I then tried to take a photograph, and even though everyone else was allowed, the Senator said that no, I would not be allowed to take a photo because I was not media. I said I was media, that I was writing articles for the Yukon News, and also books and was working on a documentary.

I sat in the front of the meeting. The Senator said that the Alaska National Interest Lands Conservation Act that passed in 1980 stated that the Coastal Plain would be opened to oil development after study. I began to shake my head in opposition. That legislation in fact said the area should either remain as wilderness, or be opened to oil development. The law left that decision to a future Congress, after studying the impacts drilling would have. The Senator then stopped his talk and said in a stern voice, "I will have you removed from this meeting if you continued to shake your head. If you shake your head again, I will have you removed from this meeting."

That was blatant intimidation. While I did stop shaking my head, because I wanted to cover the meeting and not be forcibly removed, fearing the Secret Service would harm my video camera and 25 hours of footage for my Caribou People film, the order was very wrong. It showed the tyrannical and abusive nature of the former Senator who did not understand the democratic process of the United States, instead acting as a dictator that everyone must follow.

Stevens spouted misinformation about how the residents of Kaktovik were supposedly in favor of development. To prove him wrong, Robert Thompson presented the former Senator with a petition signed by 60 Kaktovik residents opposing development. With 98 votes cast in the 2004 election, the petition was a majority of the voting public in the small village. Thompson proved that the oil industry proponents' argument that the Inupiat in Kaktovik are in favor of oil development is not true.

After the meeting was over, I said the Arctic National Wildlife Refuge needs to be protected, and we need to save the bowhead whale (which would be harmed by the proposed offshore development). I was then forcibly pushed out of the meeting by Secret Service Agents and the same police officer that I had seen at the airstrip.

Stevens has long been the leader of the crusade to destroy the last corner of North America, telling Congress that voting against drilling was voting against him, and that he would "never forget." Stevens testified to Congress in March, 2005 that he was clinically depressed over his efforts to drill in the refuge.

In 2008, Stevens was convicted of seven felonies for accepting oil industry gifts for political favors. The convictions were overturned on April 7, 2009 when a federal judge found that prosecutors had engaged in misconduct.

Oil consulting firm Veco hired contractors that more than doubled the size of Stevens' home in 2000. On July 30, 2007, FBI agents arrived with a locksmith at about noon at Stevens' Girdwood, Alaska home, near the base of the ALYESKA ski resort. Half a dozen federal agents parked at the house, with about 10 more federal vehicles parking a few blocks away at the ski resort.

They had the blinds down while they searched the house, and one carried out a black trash bag of stuff. One FBI agent told media that they were investigating whether Stevens' paid for the work, and whether it was a bribe. The IRS is investigated whether Stevens accepted monetary benefits without paying taxes on them.

Ted Stevens' son, Ben Stevens, allegedly received $284,000 for phony consulting fees from Oil industry consulting firm Veco while he was the state senate president. Four former Alaskan legislators are also charged with being bribed by the oil industry. Veco CEO Bill Allen, who oversaw the work on Ted Stevens' house, plead guilty in May, 2007 to bribing state legislators.

Alaska's only Representative, Don Young, is also under investigation. Veco allegedly held "pig roast" fundraisers for the representative, and gave Young $157,000 through their employees and political action committee between 1996 and 2006.

Young talks like an oil company. On May 18, 2006 he lied to the House of Representatives in a successful effort to oppose a global warming resolution that had passed the Senate. Young said that other countries burn as many barrels of oil as the United States. In fact, the U.S. burned 20.6 million barrels of oil in 2005, while the second largest user, China, burned less than 7 million barrels. Young also said polar bears are increasing in numbers, another mistruth that is well documented in this book.

Chapter 24

Dwindling sea ice

After being pushed out of the Ted Stevens meeting, I bid farewell to the Appalachian State College group, and decided to stay several days before heading to Prudhoe Bay. I befriended Stacey Fritz, who was working on her PhD about the Distant Early Warning line, a system to detect ballistic missiles coming from Russia.

Stacey knew Robert Thompson, whom I had met lobbying in DC. She suggested that we walk over to his house. Thompson and his wife, Jane invited us into their home.

Robert showed us an L.A. Times story of an encounter that he had with a grizzly bear while guiding a group down the Hulahula River. The bear had killed two campers, and had eaten much of them. When Thompson's raft went by the camp site, the bear charged after his raft. The bear swam fast, and Thompson guided the raft around a boulder. The bear hit the boulder, and the group got away. I was horrified at the story.

Thompson offered to take us fishing. We drove out to his small skiff. Boating in the Arctic Ocean, I was struck by the fact that I could not see a speck of ice anywhere. In 1991, also in August, ice was about a hundred feet from the shore of Barter Island. Now, the ocean was choppy, the waves forming from the winds blowing across more than a hundred miles of open water. As could be expected, it was also quite cold. The life preserver that I put on was wet. I should have had warmer clothing, but I already had an overweight pack with all of the video equipment that I brought.

The cold came from the splashing water that is inevitable with a small boat on a choppy ocean. Waves of any significance were once rare, with only small areas of open water to build up waves. To see the vast expanse of open water, where just 14 years prior the area had been near entirely pack ice was quite a wake-up call to me. I knew about the changes from following the science, but to see it firsthand, and realize the implications of this change motivated me to do whatever I could to combat climate change.

By 2007 only half as much ice remained in the Arctic Ocean as there was in 2004 at the same time of year. Such a massive change startled scientists, and was repeated in 2008. The pack ice is critical to the marine environment of the Arctic Ocean.

"The meeting of ice and open ocean habitats creates an unmatched richness of biodiversity of marine mammals," Greenpeace stated in its Answers from the Ice Edge report, June 1998. "Life concentrates at the ice edge, and its productivity each spring attracts an array of birds and mammals. Marine mammals from temperate climates venture north to feed on the seasonal abundance of Arctic waters. For other Arctic wildlife such as seal and walrus, ice provides transportation as well as a floating platform for resting, feeding and producing their young.

"Each winter Arctic pack ice advances across the Chukchi Sea and into one of the most productive marine ecosystems on Earth —

the Bering Sea. With the pack ice come animals that depend on it for life - bowhead and beluga whales, walrus, seal and polar bear. At this interface of ice and open water life flourishes and the diversity of species is greatest. The richness of this marine life in turn attracts animals and people from the land. For thousands of years indigenous people of Alaska have relied on marine resources for food, clothing and cultural traditions.

"Climate change upsets the dynamics of marine and coastal ecosystems and Native cultures that depend on them. The consequences of global warming are affecting the subsistence way of life of Alaska's Native peoples now."

"The pack ice is at its largest in April. It retreats up to 1,700 kilometers, reaching its minimum area in September. The Arctic ice is about twice the area at its maximum compared to its minimum. The advance and retreat takes the ice-associated animals to new feeding areas.

"The annual drift of the pack ice is clockwise (anti-cyclonic) in the Beaufort Sea of the Canadian Basin, often referred to as the Beaufort Gyre or the Transpolar Drift Stream. The sea ice moves away from the Siberian coast, across the North Pole and exits the Fram Strait between Greenland and Svalbard. The freshwater current in the form of ice is twice as large as the combined discharge of the four largest rivers in North America.

"The edge of the ice is not a sheer wall. It is a complex and ever-changing array of moving chunks of ice. It contains open water ponds called polynyas and long, linear cracks called leads. Walrus and ringed, ribbon, bearded and spotted seals use ice as a platform to rest, mate and molt. Ringed seal are the northernmost seal, living not just in floes of the pack ice edge, but also in the thick ice of the far north. They mainly eat Arctic cod and are the principal food of the polar bear.

"In winter, tiny marine plants grow on the bottom of the ice. These plants form the base of the Arctic Ocean food chain. Crustaceans and other small animals feed on the algae. Arctic cod and other fish eat the crustaceans, with mouths that open forward and upward to feed off the floating ice.

"As the pack ice retreats each spring, the algae growth on the ice helps to seed the phytoplankton bloom (tiny floating plants) in the open Arctic Ocean waters. Fresh water from the melting ice floes forms a clear layer above the denser, darker sea water that allows enough sunshine to produce abundant plankton. This plankton sinks after the ice retreats, feeding the bottom-dwelling or benthic communities such as clams, mollusks, worms and amphipods. Their abundance is so great that Gray whale migrate all the way from Baja California to feed along the retreating ice of the Bering Straight.

Arctic sea ice has reduced by approximately 7.8 percent per decade between 1953 and 2006 according to the report "Arctic Ice Cap Melting Much Faster than Previously Estimated," published in the Geophysical Research Letters April, 2007. That study, with scientists from the National Center for Atmospheric Research, found that Arctic sea ice is melting more than three times faster than previously calculated. Between the 1960s and the mid 1990s, the average thickness in sea ice dropped from 3.1 meters to 1.8 meters, a decline of nearly 40 percent in less than 30 years according to a Worldwatch Institute news brief March 6, 2000.

"The direct evidence is from satellite data," Climatologist Dr. Josefino Comiso said of the massive loss of ice in the Arctic. "We measure the surface temperature. Almost every kilometer of the Arctic is measured by the satellite, day in and day out. So we have this continuous record that then tells you which places are warming up and which places are cooling, just using trend analysis on this data.

"We have a different sensor that monitors the ice cover. This sensor is in the microwave region, where the wavelengths are low enough to be able to penetrate clouds without being affected too much. So this sensor is a day and night continuous sensor to afford very detailed and consistent measurements of the surface. We have 25-30 years of satellite data. We then try to extrapolate using that compared to surface data. Surface data is collected from meteorological stations all over the globe. We are consistent.

In the Arctic there are not as many stations. It is very difficult to get measurements on areas covered with ice, because the areas covered by ice are not very easily accessed. Sometimes they freeze up ships in the ice pack and leave it there for a year with scientists on board a ship. And then the scientists can get out every now and then and take measurements." The scientists' boat is frozen into the ice "so they can take measurements of the meteorological data to help us understand what is going on in the Arctic region. This is one measurement from an area that is pretty big. So it is hard to say that one measurement can represent the Arctic."

Satellite data has enabled Comiso to measure the loss of pack ice. "The advantage of using satellite data is you get global coverage, and very specially detailed coverage over the whole Arctic region. What we know for sure is the behavior of the ice cover adjacent to Alaska. We have a very detailed data set from satellites over the last 25-30 years and it used to be that in the summer, the ice pack was pretty much accessible from the coastlines of Alaska, you know you could see it from the shore. But in recent times, the ice has retreated so much that it takes a major effort to get to the ice pack from the coastline of Alaska. In the Beaufort Sea in Alaska, the ice is retreating. The Beaufort Sea forms the northern border of Alaska."

"What we have observed is over the last five years, starting in 1997, there was a little recovery in 2001, but in 2002 and 2003 the retreat was the worst. In fact you have the largest open water ever in the area adjacent to Alaska in the Arctic region. So in terms of the sea ice cover, a lot of things are really changing in the region." Previously, the pack ice rarely retreated more than 15 miles away from shore. Satellite data reveals that that gap between the north shore of Alaska and the pack ice is now hundreds of miles in

September, with major ramifications to wildlife that depend upon the ice edge to feed, rest and hunt.

Retired US Fish and Wildlife Service Arctic National Wildlife Refuge biologist Fran Mauer said, "there seems to be an increasing frequency of years when the ice was a long ways offshore. In recent years it has been quite far off shore, over a hundred miles away. The coastal erosion phenomenon has increased from the open water."

Alaskan Senator Ted Stevens reported a decrease in Alaskan ice by eight inches in 2003 compared to 2002. Submarine sonar measurements confirmed that average Arctic ice thickness in late summer since the 1950s have shown a decline to 5.9 feet in 2000 from 10.2 feet, a 42 percent decrease. The normally year-round ice is shrinking twice as fast as the overall winter perimeter, which has further alarmed scientists, and has several predicting that the Arctic Ocean will be ice-free in 5-10 years.

Like permafrost melting, the melting pack ice causes further warming of the globe. Arctic ice is white, and reflects 80-90 percent of the sun's heat back into space. Open water is much darker, and absorbs 80-90 percent of the sun's energy. The more that the sea ice retreats, the warmer the ocean becomes. The increase in warming is called a positive feedback and the potential for increased warming alarms scientists.

Harley Sorenson, in an article, "A Warning From The North Pole," in the San Francisco Examiner on August 22, 2000 wrote, "The thick ice that covers the Arctic Ocean at the North Pole has melted, leaving a mile-wide stretch of water at the top of the world, The New York Times reported Saturday." The Harvard scientist who reported the open water at the North Pole made the same trip six years earlier and that time saw ice six to nine feet deep.

"Associated with the temperature increases, we have seen a lot of impacts on snow and ice," Guntner Weller said. "The snow and ice are also good indicators of change. We have seen a fairly dramatic decrease in sea ice in the west and in the Arctic Ocean. In recent years the rate of loss has increased. Last year was the lowest (amount of ice) we have ever seen.

"These are some of the physical impacts occurring in Alaska. Associated with that is a lot of societal impacts. The native villages rely very heavily on subsistence. Marine mammals: seal, walrus and whales are affected by the change in the sea ice. Since many of these sea mammals rely on the sea ice to lie out, and for procreation, the decrease in ice is having a negative effect on them. Walrus must lie in shallow water to dive. The disappearance of sea ice can be equated to a loss or displacement of the sea mammal habitat.

Speakers at a conference in Reykjavik, Iceland in November 2004 reported that Greenland hunters had to kill their own sled dogs that year because they were unable to feed them, as the late ice build-up meant the hunters were unable to hunt seals, polar bears and walrus.

Inuit report that drowning accidents are more common, as once peaceful waters have turned into torrential rivers because of melting glaciers.

"The thickness is measured from sonar records from nuclear submarines," Weller continued. "Declassified Navy data looks at thickness reduction. From the early 50s there are a lot of U.S. and some British nuclear submarine data. They have found a reduction in thickness of 40%, which is enormous. The average thickness was 9-12 feet. When they did it recently, we are down to 7 to 8 feet, which is really quite an enormous decrease.

"In the marginal fields, on average it was six feet thick; the reduction is almost the same, so we are down to three-four feet. There is a very enormous change in the observed thickness that has been documented. It could in part be displacements by storms, shifts in ice from one area to another, but it seems very likely that most of it is due to melting of the ice rather than shifting from one corner to the next."

Weller said the location and thickness of the ice in any detail is a classified military secret, which has hindered the study of climate change in the Arctic. "The navy did not want to release the precise locations of the data. There is a lot more submarine data out there that has not been declassified, and there is a lot of Russian data that has not been declassified.

"On the underside of the ice there is a fairly thick phytoplankton bloom in spring, and the zooplankton depends on that. You remove the ice, you remove the phytoplankton mat, and that certainly could affect the whole food chain."

A report by Cynthia Tynan and Douglas DeMaster for the Arctic Institute of North America published in December 1997 found that ice-associated seals rely on the pack ice for resting, pupping and molting, and "may be especially vulnerable" to the receding ice. The report also warned that the receding ice could affect the density and distribution of ice-associated prey of marine mammals such as arctic cod and sympagic (with ice) amphipods. Arctic cod have disappeared from the west coast of Greenland around the island of Disko where temperatures have risen about four degrees Celsius over the past 15-20 years.

In his interviews with indigenous peoples, Dan Ritzman said, "It was so much about the pack ice because that is so critical to their subsistence. From the indigenous peoples' point of view, it was more about their ability to get out and hunt. The ice is a platform for them to go out and hunt. If the ice is not there, they could not go out to get food. That is what was limiting their hunting. In Wainwright and Barrow, because the pack ice was receding further, the whales were farther north, so it was more dangerous to go out to get them."

Chapter 25

Walrus

In the summer of 2007, for the first time in history, tens of thousands of walrus were unable to reach the pack ice – which was 700 miles from where it had been ten years prior. They turned around, and swam back to land. The walrus climbed up on land by the tens of thousands in Russia, resulting in the death of thousands that were trampled in stampedes. Six thousand walrus came ashore in Northwest Alaska in October, 2007.

The walrus should have been out on the ice, feeding and nursing their young. But the record warmth melted the ice about 700 miles farther than the average 1979-2000 melt, forcing walrus onto land where they had no food and were vulnerable to predators.

Researchers aboard the U.S. Coast Guard ship The Healy documented the first case of walrus pups swimming far from shore without their mothers in 2004. The scientists believe that the mothers had to leave to follow the retreating ice, which occurred weeks earlier than before. They foretold of a major population collapse if walrus cannot adapt to their apparent need to have pack ice to rest upon when nursing their calves. The pups rest on the ice when their mothers swim down for more food. Walrus mothers normally stay with their calves for two years.

The walrus in the Bering Strait, along with elsewhere around the Arctic, are in decline in both average weight and in their population.

Most dramatic is a sharp loss in young walrus. Walrus along with whales are moving northward in the northern hemisphere to keep up with the changing climate.

Recent studies by the University of Alaska at Fairbanks found that the action of walrus sucking out clams from the ocean bottom stirs and aerates sediment that is critical to the thriving fisheries there. As goes the walrus, goes the fish. The chain reaction threatens to devastate the whole Arctic ecosystem.

Walrus are very social animals, often clinging to each other with their flippers. They were once hunted to near extinction, but have since largely recovered. Walrus use their tusks to walk on the ice. In the water, they use whiskers to feel for clams, their main diet. They need thousands of clams a day, leaving large areas strewn with shells that are stripped of their meat.

"The science says the loss of pack ice will harm the numbers of walrus," Dan Ritzman said. "Walrus rest on the pack ice and feed on the bottom. The Arctic Ocean is fairly shallow, but the farther north you go, it drops off deeper. With the receding pack ice, it is getting too deep for the walrus to be able to rest on the pack ice and dive to the bottom to feed."

Walrus will break through the ice, then anchor themselves with their tusks on the ice while they rest and breathe. Inuit are allowed to hunt three walrus per year. They shoot the walrus then use a harpoon to pull the animal near. They secure more harpoons. Several other hunters abandon their hunts and come to the gunshot to help haul the walrus up, then carve up the carcass. Almost nothing is wasted. Inuit will finish by washing their boots in the ocean, teetering over the ice edge, with one hand using a long knife for the stability lost with the foot being washed in the ocean.

Walrus will frequently hug their babies with their flippers. Other walrus will sometimes help a mother and calf. In their melting habitat, often they get too crowded on the shrinking patches of ice. Walrus will display their tusks menacingly. When two evenly matched males won't budge, a fight will ensue. They charge at each other with tusks. Shrinking ice forces them to fight more for the dwindling habitat.

Walrus have air patches in their cheeks that help them float. They can sleep afloat with their cheeks inflated. Fifteen million years ago, walrus predecessors went into the sea. Now, they gather on rocky islands in the spring to rear their young in the protection of a herd.

Walrus have a hard time staying cool when hauled out on the rocks.

Alan Springer, a University of Alaska at Fairbanks wildlife specialist, said that the quarter million walrus in Alaska appear to have suffered weight loss and stress from the decrease of pack ice that they need for resting and raising their pups.

One year, walrus climbed high on a hill, then tried to take a shortcut to the sea. They followed a herd that went down a hill too steep for their ability. They tumbled down and over a bluff, one by one, crashing into a mound of dying bodies. Sixty walrus died that year. Similarly, past U.S. foreign policy of opposition to the Kyoto Protocol has pushing our global environment to the edge of no return. It is up to us to change to a saner policy of survival, before it is too late.

Chapter 26

Arctic Screaming

"The Arctic is Screaming," The Associated Press reported in December, 2007. 2007 saw a record Arctic ice melt of such magnitude that NASA scientists now predict a near ice-free Arctic Ocean as early as 2012. This could spell the demise of the polar bear far faster than scientists predicted earlier in 2007.

In 2007, the surface area of sea ice was nearly 23 percent below the previous record, dwindling to 1.15 million square miles.

"In 2007, we had off the charts warming," said Michael Steele, an oceanographer with the University of Washington. The increased size of ice melt from 2006 to 2007 was the size of the state of Alaska.

NASA warned that the ice that remained was much thinner, as well as being less in extent. Fifty scientists from 10 nations aboard the research vessel Polarstern found in 2007 that large tracts of Arctic sea ice are now only three feet deep, half the thickness of six years ago.

Reuters reported March 18, 2008 that there was a 50 percent loss of perennial (multi-year) ice between February 2007 and February 2008. More than 70 percent of the ice cap is now only one year old: 3-5 feet thick. Just a few years ago, most of it was perennial ice 12-15 feet thick.

Arctic Screaming

In 1989, 80% of the ice in the Arctic Ocean was at least 10 years old, reported University of Washington Climatologist Dr. Ignatius Rigor. Now, only three percent of the ice is that old.

The incredible scale of change, shown by the 2007 summer's record Arctic melt, needs to trigger a similar scale of change in human society to reduce the emissions causing the problem. We must meet nature's deadline, rather than any political deadline, meaning we need to move extraordinarily fast to avoid triggering unstoppable climate change.

The snow and ice cover in the summer Arctic is critical for reflecting the 24-hour solar radiation back to space. As it melts, the open water and land absorb the sun's energy, causing much more warming.

"There truly is a catastrophe brewing in the Arctic," said Ted Scambos of the National Snow and Ice Data Center. "We can anticipate that there will be very large and more rapid changes in the areas surrounding the Arctic Ocean, but more generally throughout the Northern Hemisphere. It's going to feel as though someone stepped on the gas pedal in terms of global warming."

Areas of open Arctic Ocean water were 12-14 degrees Celsius warmer than average in the summer of 2007. "We are climbing rapidly out of mankind's safe zone into new territory, and we have no idea if we can live in it," said IPCC member Robert Corell, a U.S.-based scientist (The Guardian, 2007). 2008 marked an unprecedented seventh consecutive year of much less sea ice than normal.

In December, 2007, the Center for Biological Diversity filed a petition to list the ribbon seal as threatened or endangered because of global warming. The National Oceanic and Atmospheric Administration exercised its prerogative to broaden the scope of the review to include three additional ice seal species, since they share similar habitat and biological requirements. In February, 2008 the Walrus was petitioned to be added to the Endangered Species List. In May of 2008, the polar bear was listed as threatened.

In Greenland, thousands of new rivers are forming from its melting ice sheet, and rampaging through moulins (holes through the ice). The water flows all the way to the bottom of the ice sheet, then flows underneath the ice to the sea. This is lubricating the flow of the entire ice sheet into the ocean, with enough ice in Greenland to raise sea levels 23 feet.

Preliminary satellite data found that 552 billion tons of ice melted from Greenland's ice sheet in 2007, nearly 19 billion more than the previous record. The loss of Greenland's surface ice nearly quadrupled in the last 15 years.

Scientists found more than a doubling of the rate of melting and thinning of glaciers between the years 2004-2005 and 2005-2006 in 30 glaciers in nine mountain ranges in Antarctica, Asia, Europe, Latin America, North America and the Pacific (United Nations Environmental Program, March 16, 2008).

Meanwhile, the main cause of the climate crisis: carbon dioxide, is increasing faster than ever before. In the 1990s, carbon dioxide emissions rose less than one percent per year. Between 2000 and 2005, that rate increased to 2.5 percent per year. In 2007 it was a record 2.9 percent, eclipsed only by 2008's 3.5 percent increase. Corresponding to this rise is an accelerating increase of carbon dioxide in the atmosphere, now at 2.6 parts per million per year. The atmospheric composition was 386 parts per million in early 2009. Senior NASA Scientist James Hansen warns that we should reduce levels to 350 parts per million to avoid catastrophic change.

Chapter 27

The Fate of the Polar Bear

Boating the choppy waters of the Arctic Ocean, the reality of climate change was as clear as the endless expanse of open water, without a speck of ice in sight. Inupiat guide Robert Thompson stopped by one of his fishing nets. We hauled in the net, and pulled off many whitefish, more than a foot in length, putting them into the bucket. He pointed to a large paw print of a polar bear, and advised us to look around, that they knew what fisherman were pulling in. "They are different than before – much hungrier because they cannot get out onto the ice to feed," he said. "They used to stay away from us, but now they are desperate and dangerous."

The polar bear is the symbol of the Arctic wilderness. The majesty and charisma of the Arctic giant makes it a perfect mascot for those working to combat the climate crisis. The polar bear is the canary in the coal mine for climate change, and it is time for action. The prediction of this species' extinction as early as 2040 should send a wake-up call that we must stop global warming.

The polar bear evolved from the grizzly, turning all white to blend in with the ice of the Arctic Ocean. But that ice is melting, threatening the survival of the world's northernmost and largest bear.

"It's time to put the polar bear on the endangered species list, and give them a fighting chance to survive," said Senator John Kerry.

The U.S. Fish and Wildlife Service recommended that the polar bear be listed as threatened on December 27, 2006, saying that climate change is their primary threat. That 154-page report quotes scientists' findings that polar bears are dependent upon sea ice and snow cover for survival. They then show that sea ice is diminishing rapidly with rising temperatures from global warming.

The agency found that previous projections on climate change underestimated the severity of the warming, and the speed of the melting sea ice. The USFWS quoted numerous scientists saying that without sea ice, the polar bear would go extinct.

The polar bear was proposed to be listed as threatened or endangered in December 2006, with an unprecedented upsurge in public support for such listing in 2007. The Bush administration was supposed to issue a ruling on January 9, 2008. Instead, in April, 2008 they had to be ordered by court action to issue a decision. The delay was apparently to allow the leasing of millions of acres of prime polar bear habitat in the Chukchi Sea to oil companies, without the additional regulatory hurdles that listing would cause.

"If the polar bear is to survive as global warming melts the Arctic, we need to protect its critical habitat, not auction it off to oil companies," said Kassie Siegel, Center for Biological Diversity Climate Program Director. Siegel said that the Bush Administration "seems more interested in generating oil-company profits than in complying with the law and protecting the polar bear. The polar bear cannot wait much longer. As our greenhouse emissions increase and the Arctic sea ice melts, the window of opportunity to save the polar bear is closing rapidly. Further delay in protecting the polar bear under the Endangered Species Act is inexcusable."

With the 2007 record ice melt, and NASA scientists warning of near ice-free conditions as early as 2012, urgent action is needed if we are to save the polar bear. Arctic sea ice was 300,000 square kilometers smaller in 2007 compared to 2005. The fact that we are measuring significant change in a matter of years rather than decades is something few scientists thought possible just a few years ago.

The ramifications should send all of us into the streets with signs demanding massive change extraordinarily fast to cut emissions. NASA Senior Scientist James Hansen said there is a chance to save some of the Arctic Ice, and with it the polar bear, if immediate action is taken to greatly reduce the greenhouse gas emissions that most harm the Arctic: methane and soot pollution. The Bush administration, however, tried to censor this message.

The polar bear are organized in 19 population groups across the Arctic region. They do not just roam randomly as previously thought, but go back to the same areas to hunt and den, even though they can be a thousand miles apart. The USFWS found that all 19 population groups are expected to decline based on how the decreasing pack ice effects that population group's range.

The more southerly of the population groups, such as the Hudson Bay population in Canada and the Beaufort Sea population that dens in northern Alaska are in decline because the sea ice has retreated hundreds of miles and the period in which the Hudson Bay and Beaufort Sea is thawed has dramatically increased. It is these times of no sea ice near shore when the bears go hungry. Many attempt to swim out to where sea ice had been for thousands of years, finding only open water, and drowning.

Others are resorting to cannibalism. Polar bear scientists published a paper in the journal Polar Biology in 2006 stating that they have their first evidence of polar bears eating cubs. "During 24 years of research on polar bears in the southern Beaufort Sea region of northern Alaska, and 34 years in northwestern Canada, we have not seen other incidents of polar bears stalking, killing and eating other polar bears." The scientists conclude that climate change-caused nutritional stress was likely to blame.

The Hudson Bay population declined 22 percent between 1987 and 2004, from 1,194 to 935. The USFWS report says this trend is caused by progressively earlier sea ice breakup because of global warming. The Beaufort Sea population declined from an estimated 1,800 in the late 1980s to 1,526 in 2006. The average cub survival rate declined from an average of 65% during the time period 1967 to 1989 to 43% from 1990 to 2006.

Polar bear have evolved to live on the ice of the Arctic Ocean. They are inefficient moving on land, expending about twice the average energy as other mammals when walking. In the ocean, polar bear tend to swim against the clockwise ocean currents to maintain a position near preferred habitat in their range. With ice thickness diminishing, there is an increased speed of floating ice with the oceanic current, requiring more energy and effort for the polar bears to maintain their position.

The most dramatic change is in the proportion of polar bear that build their maternal dens on land compared to on the sea ice. Historically, 90 percent of polar bear dens have been on sea ice. With the ice melting, from 1985 to 1995, polar bears hibernated on land an average of 36.6 percent of the time, and the rest of the time out on the ice of the Arctic Ocean, according to world renowned polar bear expert Dr. Steven Amstrup (USGS). But with pack ice having reduced in thickness, new findings show that polar bears made their dens on land 63.6 percent of the time between 1996 and 2005. Armstrup said that there is still time to save the polar bear on the Canadian Archipelago if we significantly decrease greenhouse gas emissions.

The Arctic National Wildlife Refuge coastal plain is a critical place for polar bear to den in North America. Polar bear make their maternity dens on the refuge's coastal plain in much greater density than any other area on the North Slope of Alaska. Already 95 percent of the North Slope of Alaska is potentially opened to oil development. The last 5 percent is the coastal plain of the Arctic National Wildlife Refuge. With bears now using land-based dens more than twice as often as just two decades ago, climate change has doubled the importance of protecting the refuge for the survival of the polar bear.

As if being pushed onto land was not enough, the dens themselves are collapsing because of warming winter temperatures. Climate change has caused earlier spring snowmelts that have collapsed the dens of both polar bear and their prey, ringed seals, killing cubs and pups during their most sensitive time period (USFWS, 2006).

The USFWS reported that climate change is impacting the quality and quantity of snow for denning. Without enough snow for den construction in the global warming Arctic, polar bears are searching harder for den sites, and have to build inferior dens. Then, the earlier snowmelt and rainstorms during the winter cause dens to collapse while the cubs are still immature, and they die within days.

The most critical factor for polar bear reproductive success and cub survival is the availability of ringed seal pups from mid-April to ice-breakup, currently in July. The decline in sea ice is expected to cause a corresponding loss of ringed seal, and with that the polar bear as well.

The Endangered Species Act is unequivocal in its mandate for measures to save species from extinction, whenever it is humanly possible. Scientists say it is possible to save the polar bear with major greenhouse gas reductions. The law now mandates this reduction. It just needs to be enforced.

The Bush administration suppressed the science on polar bears and global warming for years, said Karen Suckling, policy director for the Center for Biological Diversity. "They're deeply afraid the Endangered Species Act will create a clear regulatory requirement to limit greenhouse gas emissions," she said (Washington Post, April 15, 2007). President Barack Obama is a refreshing change, but he needs to move faster to make up for lost time.

Guntner Weller warned that an ice-free Arctic Ocean "would spell a disaster for polar bears. We had a situation last fall, when the freeze-up occurred late and the polar bears could not get out on the ice. There was an accumulation of polar bears at Barrow that has never been seen before. They became a nuisance."

Polar bears have suffered a 15 percent decrease in their number of offspring and a similar decline in weight over the past 25 years, according to the Arctic Climate Impact Assessment Council, an international commission of eight countries, including the US and six indigenous groups in a November, 2004 report. The report was written by a team of 300 scientists. With predictions of an ice-free ocean later this century, the Arctic report questions whether the polar bear will survive in the wild.

The World Wildlife Fund (WWF) found that in the southern range of polar bears, for example the Hudson and James Bays of Canada, sea ice is now melting earlier in the spring and forming later in the autumn. The time bears have on the ice, storing up energy for the summer and autumn when there is little available food, is becoming shorter. As the periods without food become longer, the overall body condition of the polar bears declines. This is particularly serious for bears that are pregnant or have cubs, and for the cubs themselves. In Hudson Bay, scientists found the main cause of death for cubs is either lack of food or lack of fat on nursing mothers.

For every week earlier the ice breaks up in Hudson Bay, bears come ashore roughly 10 kg (22 lbs) lighter and in poorer condition according to WWF. Reduced body condition can lead to lower reproduction rates, which in the long run could lead to local extinction.

"Polar bears are the largest land predators in the world," Deborah Williams said. "They are extraordinary animals that have evolved and adapted magnificently to their challenging environment. And polar bears need ice. When you deprive polar bears of ice and they get stranded on land, several things happen. One, they don't have their food source: seals. Two, they are more likely to be killed, 'in defense of life and property.' So you have a lot more human-caused mortality of polar bears. And three, polar bears become more engaged in fighting amongst themselves."

In addition to a loss of habitat, the massive reduction in ice extent also reduces the overall food levels in the Arctic Ocean, scientists warn. Dr. Vera Alexander of the Fisheries and Ocean Science Department at the University of Alaska, has been observing dramatic changes in temperature and ice-edge extent for the past 20 years. An expert on ice-edge ecology, Alexander stated that the continued decline of the sea ice will affect the production of algae, which live beneath the ice, and form the very base of the arctic food chain. "Without the ice algae," Alexander noted, "there would be no possibility of a food chain as we know it." The resulting impacts would ripple up the food web affecting fish, seals, whales and polar bears.

Dr. Josephino Comiso said that "sea ice is a habitat for polar bear and many other animals. The sea ice is not a permanent habitat. It is a floating entity, on top of the ocean. You would think that polar bear have stable habitat. But as this ice retreats, then this platform that the polar bear uses to find food will not be available anymore."

Retired U.S. Fish and Wildlife Service Arctic National Wildlife Refuge biologist Fran Mauer said, "The numbers of bears (stranded on shore) have been increasing compared to what people used to see. During summertime it was an unusual event to see a polar bear along the shoreline. Not totally unheard of, but unusual. Now in late summer in the last few years, the numbers of bears and observations of bears has dramatically increased." Mauer said the unusual congregation of bears has been found around Barrow and Kaktovik, feeding on the carcasses of whales that are hunted by the Inupiat.

Once considered a nuisance, people would set traps with bait and a gun that fired when the bear stuck its head inside. Now, we attack the bears with pollution, oil industry development and climate change, as well as guns. Pollutants travel from up to a thousand miles away. They bio-accumulate in the top predators, like seal who eat Arctic cod. Polar bear-eating seal get the most toxins. Polar bears have high concentrations of PCBs in the mother's milk. Those with the highest levels also have the highest rate of cub mortality.

But climate change has eclipsed PCBs as the primary threat. Deborah Williams said, "I am personally desperately worried about the plight of the polar bear."

The December 27, 2006 USFWS report recommending that polar bears be listed as threatened concluded that the "future persistence of polar bears is tenuous," and "ultimately, if sea ice disappeared altogether, polar bears would become extinct."

A 2004 study found 4 polar bear dead as they tried to swim from the shore of northern Alaska to the distant pack ice. Extrapolating the data to the entire population, because only 11 percent of it was surveyed, the scientists found that 27 polar bear likely drowned during this one event. Larger areas of open water are increasing wave size, which also contributes to polar bear drowning.

In a speech in Anchorage on September 24, 2006, Dr. Armstrop reported finding polar bears dead in a severely emaciated state. Even the bone fat was gone in some of the polar bears he examined, which occurs in the final stage of starvation. It will take massive reductions in greenhouse gas emissions, aerosols and soot to give the polar bear a fighting chance of survival.

Around the country, groups are establishing the polar bear as their mascot. The majestic creature is such a charismatic victim that hopefully can help reach people's hearts and minds to make the needed changes.

*

After returning with a bucket of fish from Robert Thompson's net in the Arctic Ocean, we relaxed at Thompson's house and had a delectable caribou dinner. Later, Stacey and I walked the beach of Barter Island, getting wood for a fire in the bunk house that we shared. The next day, we walked past the air strip, about a mile from where we stayed. About a mile farther, we reached the whale bone yard. The Inupiat people hunt two Bowhead whale per year to feed their people and sustain their culture. Many oppose this hunt because bowheads are endangered species. Polar bear were known to regularly visit the area to feed off the bones. I had gone down to the bone yard several times to fish, always keeping my bear pepper spray ready. Fritz talked of traveling with her partner every year to catch hundreds of pounds of salmon which they smoked and canned.

Later, we walked past the village toward the old Distant Early Warning System barracks. We came upon an area where a known toxic waste dump was just eroding away into the Arctic Ocean, a few hundred yards from Kaktovik. These areas were formerly thought to have been "permanently" frozen into the ground: that is until the climate change of our own making. To think that this waste was oozing into the Arctic Ocean right where dozens of fishing nets are set by Inupiat is outrageous.

The cost of dealing with the mess in such a remote location has hindered significant clean-up.

The erosion was not just in Barter Island, but all throughout the Arctic Ocean coastal region, where similar toxic and radioactive dumps are now being released into waterways and the ocean.

Miller said a major change she has seen recently in Arctic Alaska is "the change due to coastal erosion, and the tremendous storms that occur along the Arctic Ocean coastline in the fall (August to October). Where in the past you had more sea ice near the shore, you can look out from the coastline of the Arctic Refuge and as far as you can see to the north is open water in September. And that is a major change. You can see where there has been coastal erosion from big storms.

"That has potential impacts for industrial infrastructure and oil field expansion. For example building buried sub-sea pipelines to the Northstar island, the question is whether they've taken into account how much that beach may erode over the 50-year lifespan of that field. There are a lot of questions about that."

The 176 million-barrel Northstar oil field is the first offshore development in Alaska's Arctic. With all of the coastal erosion of the Arctic Ocean shoreline, Greenpeace and Alaskan Natives challenged the development in court. They were unable to stop the disastrous development.

Chapter 28

The Decline of Black Guillemots

Polar bear are not the only victim of the melting polar ice. The entire ecosystem is collapsing along with the ice. Being so remote and hard to access, much of the biological devastation is happening away from the watchful eye of scientists. But the impacts of one bird – the black guillemot – is being studied by scientist George Devoky.

Black Guillemots depend upon the edge of the sea ice for food. They have orange feet. In the summer, they are all black except for a white patch on the top of their wings and a white underside of their wings. In the winter they turn mostly white. Guillemot is the common name for three species of birds in the auk family, two of which live in North America.

They breed on coasts and islands, like other auk species. Unlike other members of the auk family, they nest in dense colonies. They hide in crevices of the cliffs overlooking the Arctic Ocean, raising their young. The nests are near impenetrable for predators.

All at one time, so as to swamp predators and maximize chick survival, mother guillemots will call to their young, begging them to jump to the sea. Baby guillemot jump off the cliff and glide to the ocean. Adults will escort guillemot chicks down, because gulls will catch those that are not escorted. Those that do not jump to the sea sometimes get lost, wandering around helplessly on the cliff tops.

They are easy prey for Arctic fox and gulls. Devoky found that the guillemots he has been studying are in decline, attributable to global warming.

I met with George Divoky in Seattle, having traveled there by train, where we talked at length and I had dinner with his family. Thirty years ago, he began work at Cooper Island, located far in the Arctic Ocean, 530 kilometers north of the Arctic Circle. He discovered some black guillemots, far north of where he had expected them. Finding that nesting habitat was their primary limiting factor, he began turning over debris that the military had left, such as boxes and crates. These served as nests, and the population grew to more than 200 pairs in the 1980s.

Deborah Williams said, "He found that if he made nesting sites more birds would come and propagate. He was doing this experiment about guillemots and nesting sites and it was quite fascinating. The populations grew and grew as the nesting sites increased. What he found was then the nesting sites started declining. The decline was corresponding to how far the birds had to fly to get to the ice edge, because they were feeding on the highly productive ice edges. That is where the parents would go out to get food for their young.

"The ice edge was retreating year after year after year, and the birds had to expend more energy to fly out to the ice edge to get the little fish that they feed on; there was greater and greater nest failure." In 2003, when the ice edge was at its furthest, Divoky reported a 90 percent nest failure. Williams continued, "What you can find with global warming is in so many instances it makes feeding so much more difficult. You have increased mortality associated with that. That is a profound impact."

Divoky found that the guillemot chicks were starving, as their parents could not make it out to the ice edge to get Arctic cod and make it back, the distance being so great. The scientist also reported behavior he had never seen before, such as males choosing two females instead of one because starvation meant there were not enough mates, and chicks walking away from their nests in search of food and foster parents.

Divoky discovered that the snow was melting five days earlier each decade since his study began. The guillemots were laying their eggs five days earlier per decade as well, coinciding with the earlier melting. Divoky said, "birds (are) non-political. They have no choice but to react." That is what we, humans, as the unquestionable cause of this increased warming, need to do as well. We must work with all of our power to stop our emissions of greenhouse gases, the cause of this problem.

"You get these people who say 'Do you really think it's happening?' and I'm like, "What is it you don't understand?" Divoky said. "I just hope you care about what the guillemots are telling us. And I think that this sort of rapid rate of change—that in three decades, and even within the length of one study, to see this much change indicates that things are happening globally that humans have to take note of."

Chapter 29

Sea Coasts Crumble Into Sea

All around the circumpolar Arctic, the shoreline is crumbling into the sea as the permafrost that once held the land in place melts, and sea levels and waves increase in size.

Two villages on the Bering Strait, Shishmaref and Kivalina, have decided to move at a cost of hundreds of millions of dollars each. The General Accounting Office (GAO) found that flooding or erosion regularly affects 86 percent of the state's 213 predominantly Native villages. The GAO predicted in a 91-page report in December, 2003 that four villages, including Shishmaref and Kivalina, are in imminent danger. Newtok, where the Ninglick River erodes 90 feet of bank a year will have to move in five years according to the report, and Koyukuk must move soon as well.

Five other villages are threatened by the erosion. Kaktovik (293 residents), Point Hope (population 757), Unalakleet (with 747 people) and the regional hubs of Barrow (population 4,581) and Bethel (population 5,471) have airport runways, roads and other facilities threatened by erosion.

Stevens predicted that Barrow will also have to relocate. Its cost would be exponentially larger than the mammoth efforts necessary to move Sishmaref and Kivalina. "We face the problem of moving native villages that have been located along the Arctic and West coast of Alaska for centuries because they are slowly but surely being inundated by seawater," Stevens told five top climate scientists testifying before the Senate Commerce Committee.

"In the coastal zone, the impacts of climate change are more severe," said Guntner Weller. "In two villages, Kivalina and Shishmaref, the coastal erosion is getting so severe they are forced to move the village at a cost of a million dollars per person. The cause of this coastal erosion is three-fold. As the permafrost thaws, the coast collapses. Second, with the glaciers melting, sea levels rise, causing coastal erosion to increase. Third, more open water allows storm surges to develop stronger and more severely. The larger fetch area (open water) has caused many more frequent storm surges and severe storm surges."

"On erosion, we are fairly certain that it has been accelerated much by climate change" Weller said. "It is true that there has been erosion in the past, but the rate of it has been greatly increasing. Now there is a legal argument, as to who pays for it. Is it really the oil industry or the over-consumption of fossil fuels? I think there is good evidence that this impact is caused by global climate change. It is true that the Arctic coast is pretty dynamic. It has occurred before, but never at this magnitude."

"Global warming's impact on infrastructure has been well documented," Deborah Williams said. "It is most profound in terms of impacts on coastal communities. We now have several coastal communities that are being washed into the ocean and must be moved. Sishmaref is the most impressive, or depressing example of that. What you have with these coastal communities is generally three impacts going on at the same time. One is the permafrost on which they sit is melting. Two is that sea levels have increased. And three is that in the past when you had ice covering most of the Bering Sea and Chukchi Sea for so much of the year, particularly in the spring and fall, when the winds came, the winds did not create significant waves. But when you don't have ice — and they have fewer days of ice on the Bering Sea and Chukchi Sea — the winds create wave action and the waves erode the banks of the village.

"It is a very dramatic loss to these communities, and very expensive. It will cost a hundred million dollars at minimum to move these coastal communities. This not only has economic costs. It also has profound social costs. This is where indigenous people have lived for thousands and thousands of years, so they are losing their homelands."

"There is the village of Shishmaref which is basically washing away," Charles Wohlforth said. "The waves need room to build up. In former times, there were fewer times when there was a gap between the shore and the ice. In the fall, which is a stormy time of year, if the ice is near shore, regardless of the strength of the wind, the waves would not get that large. Now for the last two years we have record low amounts of ice extent. There is more open water than ever before. The result of that is you get these big sea waves that come crashing into the shore during storms or just during normal conditions because there is hundreds of miles of open water out there. That has very clearly caused accelerated erosion, but again it is difficult to tell exactly how much because the shore of the Arctic is not that very well mapped. It is hard to tell what was there previously."

"In Shishmaref and the Bering and Chukchi Seas, the bluffs are receding up to 30 feet per storm," Bruce Molnia said. "They put boulders on the shoreline to protect it, and all of those got washed away. There used to be shoreline ice that no longer accumulates with the warming." That shoreline ice once buffered the shore from the storm surges.

Pam Miller of Arctic Connections in Anchorage said, "Standing along the shore of the Arctic Ocean you see big ice wedges just melting into muddy little rivulets that they make going out into the Beaufort Sea. You'll have this little cliff as the facade of the tundra hangs out with a carved-out area under it. It is an overhang where it is melting out underneath it. For Alaska, there are changes. It is going to affect the infrastructure of villages, and the social investment of a lot of money."

Chapter 30

Kivalina

Kivalina is a traditional Inupiat village located at the tip of an eight-mile barrier reef, between the Chukchi Sea and Kivalina River, 80-miles northwest of Kotzebue. Subsistence hunting of whitefish, salmon, seal, walrus, whale and caribou forms the bulk of the protein of the people in Kivalina. The village has decided to relocate, in part because of the eroding beach that scientists say is being caused by global climate change.

In February, 2008 Kivalina filed a global warming lawsuit against nine oil companies, fourteen electric power companies and one coal company demanding that they pay for the damages caused to Kivalina residents, including the cost of relocation. This is the first global warming lawsuit with a "discreet identifiable victim." The lawsuit claims public nuisance under federal law, private and public nuisance under state law, civil conspiracy and concert of action. It states that there was a conspiracy to mislead the public regarding the causes and consequences of global warming.

"The reason we have to relocate is we are overcrowded on our little sand spit of an island, and we have no place to build anymore," said Kivalina Mayor Don Hawley. "There is no room to install sanitation facilities. We are surrounded on four sides with water. Our sanitation condition in this village is totally third world. We are totally lacking sanitation facilities. We have to—for the good of our people—we need off this island.

"There's been erosion from natural conditions, from the spring melt-off and also from waves from the ocean. What the ocean is doing is getting more active. As far as waves are concerned, we are getting a lot more south-southwest winds. The waves are just eating away at the beach.

"On the lagoon side of this island, there's been a few people that have houses on the northeast side of the island. They have noticed that some of the spring melt-off has been eating away at the bank and it has been slowly closing in on their houses. In the past the state and federal government have funded erosion control projects and they slowed down the pace of erosion."

Hawley said that eventually the airport will be next in line for erosion control. "If the waves keep coming like they have been, the airport will probably have to be addressed for erosion control. Once that happens, we'll seriously consider declaring an emergency."

Kivalina's average elevation is eight feet. With a population of 377, the village covers 1.9 square miles. In 1913, Northwest District Superintendent for the U.S. Department of Interior Walter Shields wrote a report on his visit to Kivalina when "the thermometer froze at—65 degrees so we do not know just how cold it was." Those temperatures that so characterized the Arctic for millennia are now near nonexistent.

One irony with global warming is snowfall often increases. This is because there is more open water that evaporates moisture into the atmosphere. More moisture means more snow.

"Within the last ten years, snowfall has been heavier because of the warmer conditions. It seems like we're getting more low pressures and lots of moisture. What's happening is either we will get rain or snow when what we used to have was a good freeze and the rivers would be frozen solid and the ocean would be frozen solid. But lately the rivers have been real unstable and also the ocean.

"The ocean is not freezing like it used to. And the rivers don't freeze like they used to. Last December, the river was free flowing. It only froze over in the last part of January. It affected the fish in the river—we couldn't find them.

"It's caused some concern, not only for the residents but also for what it has done to the animals. The animals that we rely on, instead of being in the usual places where we find them, they will move to different locations because the conditions are warmer." Global warming has totally changed the migration route of the fish and the ocean mammals, Hawley said.

"Currently the relocation (of the village) is under study. It has been that way since the relocation idea started back in the 80s. And so far all we are seeing is studies and more studies." The Kivalina escape road alone is estimated to cost 15 to 21 million dollars.

Hawley said a major storm hits the village every fall. “In October 2002 it was moderately severe and it actually washed out one of our drain-field facilities. That was the only extent of erosion that we had there. But on the airport side, the airport manager has been noticing that the beach has been eroding at the pace of one foot every storm. With global warming lately we’ve been getting the storms periodically over the summer.

“People realize that if the trend continues, the migration routes will probably change. There was one year when the spring melt-off was different. Some of the elders said this is the first time that we’ve seen the river do that ever.”

Chapter 31
Shishmaref

"The villagers of Shishmaref plugged holes in their crumbling sea wall with sandbags ... to protect exposed permafrost from further erosion. A raging storm last week caused parts of the sea wall to crash into the Chukchi Sea, leaving homes literally teetering on the edge of the ocean embankment and exposing underlying permafrost to the sea. The warmer ocean waters have melted the permafrost, speeding up erosion," the Anchorage Daily News reported on October 9, 1997. The 1,300-foot wide barrier island is losing 25-125 feet in width every year.

Shishmaref is a Native American village located on the Bering Straight on the far western coast of Alaska, just 100 miles from Siberia. The village has a population of 590, with 136 families. Archaeologists found evidence that native peoples have lived there for centuries. The original Inupiat name for the island is Kigiktaq. In 1816, Lt. Otto Von Kotzebue named the inlet Shismarev after one of his crew. After 1900, a supply center was established there to supply gold mines, and the village was renamed after the inlet. But severe erosion that scientists have linked to global warming has caused homes to collapse into the sea, forcing the emergency movement of 20 homes, the building of a $110,000 sea wall and in time, officials say, the inevitable movement of the entire village. Villagers are worried that a gradual move will split the village, or worse that lack of funding might force people to move to other villages, abandoning Shishmaref.

The U.S. Army Corps of Engineers found that it would cost 102 million dollars to move the village from the barrier island where it currently resides to a hundred-acre area on the mainland. The General Accounting office reported that the cost may be as high as $400 million, and questioned whether it was worth the cost. The Corps of Engineers requires local communities to come up with 25-50% of the costs, something impossible for the village, and creating a dilemma that policy makers will increasingly need to deal with.

On July 20, 2002, the village of Shishmaref released the unofficial count of a vote to relocate the village. The vote was 161-20 in favor of moving. In January 2004 the village picked Tin Creek, 12 miles to the south, as the future site for the village. Tin Creek is near Ear Mountain, which villagers say can provide gravel for the new buildings, roads, airport, sewage lagoons and other development.

Shishmaref Lutheran Church Pastor Kim Oslovich said, "I've only lived here for six years but there has definitely been significant erosion over the course of those six years. There's significantly less shoreline, less beach, especially on the western side of the village. I think too that there has been some weather change. Everyone says that there has been some weather change. Even in the short time that I've been here, it seems like the fall storms are more severe and freeze-up tends to be a little later, things like that.

"The most obvious impact is more than 20 homes had to be moved from one part of the island to the other, and that was a major enterprise. The way it is now, the homes that were in imminent danger have all been moved. But right now there is a sea wall that is being constructed, so hopefully that will ameliorate some of our problems at least in the short-term, but as you know, holding back the ocean is not a thing you can really do. It is just a matter of time before we start losing land again.

Shishmaref is located on a sandy barrier island on Alaska's Chukchi Sea Coast, five miles from the Seward Peninsula mainland. It is near the Bering Land Bridge where native peoples and many animals crossed into North America during the last ice age. "People have lived here for a long, long, long time," Oslivich said. "There are archeological digs that have found things that are really old, more than a thousand years old. But as far as a permanent settlement, that came into being when there was contact with European people, the Americans and Russians. And that would be in the mid to late 1800s.

"Erosion has been a consistent problem over the years. There was a particularly bad storm back in the early 70s that caused significant erosion. But I think the thing is it has definitely accelerated more in recent years, especially since 1997. We had a really bad storm in '97 that caused a huge amount of erosion, between 50 and 100 feet. That was when the first big batch of houses had to get moved. There were about 15 houses that were moved then. Since then, we have had more. Just last year we had significant erosion and three houses had to get moved.

"One of them had to have an emergency move 20 feet back because it got to the point where you would walk out the door and go right over the edge of the cliff. It was an October storm, and it really wasn't that bad of a storm," he said, estimating the wind speed at 30 knots. "It was not extreme weather, just bad enough and the wave action was with the winds the wrong way that it caused problems.

"It was dramatic. It was a dramatic loss of land. Before the storm there was a fairly wide beach. You could go down and drive on the beach on a four-wheeler. There was lots of space and people had their traditional drying racks for meat. After that the beach was just gone. The beach just disappeared. People lost a lot of meat that was on drying racks that got washed away.

"The governor declared a state of emergency. Houses were precarious on the edge of the bluff. It was pretty significant. One house that was very close to the edge that nobody was living in crashed over the bluff into the sea."

Shishmaref resident Tony Weyiouanna is working with government agencies to relocate the village. A storm in November 2003 left five boats missing, including Weyiouanna's 22-foot plywood skiff. The storm forced the evacuation of five residents. Eight-foot waves chewed away at the bank, Weyiouanna reported.

All the village lowlands were flooded and waves lapped the edges of the villages' washing facilities and sewage lagoon.

"As people have reflected on these things, they will say you know, there does seem to be a correlation," Oslovich said. "It may well be that as things get warmer, the sea levels rise, the freeze-up is later. There is more damage from storms in the fall because the water is open for a longer period of time. I think some people are starting to make that connection over the last few years.

"We need to care for creation. If we don't care for it, we are all going to suffer. You see that on the micro-scale. You see that people around here traditionally only took what they needed from the land, and they used everything that they took, and that kind of balanced things out. And on the large scale, we do need to be concerned about those larger issues. And people around here are starting to think in those terms. They are starting to think in terms of how much pollution do the snow machines produce, and what kind of effect does that have. We'll see where that takes us.

Chapter 32

Opening the Northwest Passage

Boating in the Arctic Ocean, where for millennia all had been ice, I witnessed the opening of the northwest passage in 2005. Already, non-ice breaking boats have navigated through in recent years. The passage has been open about 20 days a year, and scientists predict it will be open 100 days per year by 2050. Canada has claimed this territory as its own inland waters. The U.S. and many other countries, however, call it international waters, and do not intend to ask permission for passage.

The fabled Northwest Passage was originally sought as a trade route between Europe and China and India. Attempts to get around North America began in the 1490s with voyages by John Cabot that were unsuccessful. Cabot, with the support of many wealthy European merchants, believed there was a passage to Asia going west. When reports came back in 1493 that Christopher Columbus had (erroneously) reached Asia, Cabot began work on a direct passage through the north. In 1496 he got approval and funding from King Henry VII of England.

He set sail from Bristol on 1497, going generally northwest. After a rough crossing of the Atlantic, he sailed along the Labrador, Newfoundland and New England coasts. He thought the land was in northeastern Asia, and claimed the area for the King. He returned with a group of four ships and about 300 people. He reached

Greenland and sailed north until his crew mutinied because of the severe cold. He was forced to return to England because he was so low on supplies, and died shortly thereafter.

Sir John Franklin was a decorated war veteran for England, and had extensive experience in explorations of the Arctic regions of Canada and Alaska. In 1845 he left with a crew of 129 and two boats, the Terror and the Erebus. The ships were last seen by a whaling vessel in Baffin Bay on July 26, 1845. Ships were sent to find them from 1848 to 1859. Franklin's second wife outfitted an expedition in 1847—the Fox—that finally found the remnants of the expedition. The ships had become locked in ice in September 1846 between Victoria Island and King William Island. Franklin and others died, and the crew deserted the ships in 1848, only to die in the frigid northern lands.

The existence of the Northwest Passage was proven in 1854, and Roald Amundsen finally made the passage from the Atlantic to the Pacific from 1903 to 1906. During this expedition, he determined the position of the north magnetic pole. In 1969, the U.S. icebreaker the Manhattan became the first large ship to make the northern sea route.

The Russian ship Akademik Fyodorov was the first ship to reach the North Pole in 2005 without the help of an icebreaker. More than half a million tons of grain was shipped through Churchill, Manitoba, in 2007: once a near entirely iced-in port.

Colin Nickerson wrote in the Boston Globe, March 21, 2000 that, "The rapid retreat of the ice is also opening the way to direct challenges to Canada's control over the waters of its vast northern realms—above all, its claim to the legendary Northwest Passage." Nickerson described the area as one of the most sensitive environmental zones and a critical strategic arena. The Northwest Passage refers to numerous different passages through the northern territories.

"The opening of the northwest and northeast sea routes will allow the possibility of ship traffic through the Arctic Ocean: a distance much shorter and much more economical," Climatologist Guntner Weller said. "But this ship traffic could also increase pollution with tanker traffic."

Shipping across the Northwest Passage would devastate the Inuit peoples, warned Inupiat leader Robert Thompson. Inevitable oil spills would contaminate the fragile lands, and disrupt the marine ecosystem.

Canada's head of ice breaking operations predicted in 2000 that there was no question that the Northwest Passage would open, the question is whether it would be in a few years or decades.

The United States Ambassador to Canada, David Wilkins, said that the U.S. does not recognize Canada's claim that the Northwest Passage is Canadian waters. Canadian Conservative Prime Minister Steven Harper responded that, "we have significant plans for national defense, and for defense of our sovereignty, including

Arctic sovereignty. It is the Canadian people we get our mandate from, not the ambassador of the United States."

In August 1999, Inuit fishermen were surprised by a submarine that surfaced right next to them. By the time Canada's patrol planes reached the area, the sub had submerged. The Inuit reported more sub sightings in the next few weeks. The Canadian government thinks that the sub was likely French or American based on descriptions of it from the Inuit.

The issue has been a sore point between U.S. and Canada, with the commander of the Canadian forces of the far north saying it may need to use firepower to enforce its claims over the 1.5 million square miles of northern ocean terrain.

Chapter 33

Pentagon Calls Climate Change More Serious Threat Than Terrorism

Conflicts such as those over the opened northwest passage have caught the attention of the Pentagon, which in 2004 found that climate change is a more serious threat to national security than terrorism. Rather than heed their warnings, former President George Bush tried to censor the 2004 Pentagon Report on Climate Change.

The hundred thousand dollar Pentagon Report found that "climate change vastly eclipses the threat of terrorism," according to the Observer Guardian. The report details how climate change could result in a global catastrophe costing millions of lives in wars and "natural" disasters.

The Pentagon document reports that abrupt climate change is like the tipping of a canoe that may start relatively slowly, then plunges over in violent upheaval as the natural forces of gravity flips the boat to another state. The report warns this could bring the planet to the edge of anarchy as countries develop nuclear weapons to defend against dwindling food, water and fossil fuel energy supplies.

The threat of climate change to global stability vastly eclipses that of terrorism, the report reads. The Pentagon report states, "Disruption and conflict will be endemic features of life. Once again, warfare would define human life." The British Observer-

Guardian reported, "The findings will prove humiliating to the Bush administration, which has repeatedly denied that climate change even exists. Experts said that they also will make unsettling reading for a President who has insisted national defense is a priority."

The report was commissioned by Defense Advisor Andrew Marshall, a leading Pentagon official for the past 30 years, who is called the "yoda" of the Pentagon by insiders. It reads, climate change "should be elevated beyond a scientific debate to a US national security concern."

The Pentagon report found that, "Rather than decades or even centuries of gradual warming, recent evidence suggests the possibility that a more dire climate scenario may actually be unfolding."

Fortune Magazine said, "The threat that has riveted their attention is this: global warming, rather than causing gradual, centuries-spanning change, may be pushing the climate to a tipping point. Growing evidence suggests the ocean-atmosphere system that controls the world's climate can lurch from one state to another in less than a decade—like a canoe that's gradually tilted until suddenly it flips over. Scientists don't know how close the system is to a critical threshold. But abrupt climate change may well occur in the not-too-distant future. If it does, the need to rapidly adapt may overwhelm many societies—thereby upsetting the geopolitical balance of power."

The report also warns that climate change will make a future with the borders of the U.S. and Australia being patrolled by armies firing into waves of starving boat people desperate to find a new home. Fishing boats will be armed with cannons to drive off competitors and demands for access to water and farmland are backed with nuclear weapons. Greenpeace called the Pentagons warnings as a "coming climate Armageddon." Sir David King, Chief Scientist in former British Prime Minister Tony Blair's government, also said that global warming is a greater threat than terrorism.

The Observer reported key findings of the Pentagon Report as the following:

Future wars will be fought over survival rather than religion, ideology or national honor.

Violent storms smash the coastal barriers of Holland, rendering large parts of the country uninhabitable. The delta island levees in the Sacramento River area will be breached in California, disrupting the aqueduct system transporting water from north to south.

Realizing the threat climate change poses, an independent panel appointed by the World Bank recommended in 2004 that the institution phase out investments in fossil fuels within eight years and aggressively invest in renewable energies for the needed transition into fuels that do not cause global warming. The panel wrote, "The WBG (World Bank Group) should aggressively increase investments in renewable energies by about 20 percent annually. WBG lending should concentrate on promoting the transition to renewable energy." Unfortunately, World Bank leaders do not plan to follow their consultants' advice.

Meanwhile former President George Bush said, "The jury is still out." Greenpeace released a statement February 22, 2004 stating, "Actually, Mr. Bush, the jury's been in for some time, and now even a report commissioned by your own Pentagon is saying you're wrong. Perhaps it's time you focused on the real terrorist threat to our planet: the oil companies like Exxon which continue to fund your re-election, and whose interests you continue to defend at the expense of our future.

"While you're pursuing policies that accelerate the production of greenhouse gases and continuing to deny the existence of a threat, the World Bank is being told it has to stop subsidizing Armageddon, and the Pentagon is war-gaming ways to survive a catastrophe it's calling plausible.

"If you were willing to launch a preemptive war on enemies you believe may someday think about attacking the US, wouldn't it seem prudent to take preemptive action against climate change?"

The Pentagon warns of a climate spiraling out of control with the positive feedback loops of greater warmth and ecological catastrophe. The following is the prediction by Pentagon leaders in the 2004 report:

"As temperatures rise throughout the 20th century and into the early 2000s potent positive feedback loops kick-in, accelerating the warming from .2 degrees Fahrenheit, to .4 and eventually .5 degrees

Fahrenheit per year in some locations. As the surface warms, the hydrologic cycle (evaporation, precipitation, and runoff) accelerates, causing temperatures to rise even higher.

Water vapor, the most powerful natural greenhouse gas, traps additional heat and brings average surface air temperatures up. As evaporation increases, higher surface air temperatures cause drying in forests and grasslands, where animals graze and farmers grow grain.

As trees die and burn, forests absorb less carbon dioxide, again leading to higher surface air temperatures as well as fierce and uncontrollable forest fires. Further, warmer temperatures melt snow cover in mountains, open fields, high-latitude tundra areas, and permafrost throughout forests in cold-weather areas. With the ground absorbing more and reflecting less of the sun's rays, temperatures increase even higher."

The following are predictions of the 2004 Pentagon report:

- Severe storms in 2005, the year that a record three category five hurricanes formed in the Atlantic, and Hurricane Katrina wiped out New Orleans.
- More severe storms and typhoons that bring about higher storm surges and floods in low-lying islands such as Tarawa and Tuvalu (near New Zealand).
- Particularly severe storms causing the ocean to break through levees in the Netherlands making some coastal cities such as The Hague unlivable. Failures of the delta island levees in the

Sacramento River region in the Central Valley of California creates an inland sea and disrupts the aqueduct system transporting water from northern to southern California, because salt water can no longer be kept out of the area during the dry season.

- Melting along the Himalayan glaciers accelerates, causing some Tibetan people to relocate.
- Floating ice in the northern polar seas, which had already lost 40% of its mass from 1970 to 2003, is mostly gone during summer by 2010. As glacial ice melts, sea levels rise and as wintertime sea ice extent decreases, ocean waves increase in intensity, damaging coastal cities. Additionally millions of people are put at risk of flooding around the globe (roughly 4 times 2003 levels), and fisheries are disrupted as water temperature changes cause fish to migrate to new locations and habitats, increasing tensions over fishing rights.

Fortune Magazine reported in 2004 that the Pentagon predicted abrupt climate change "would cause massive droughts, turning farmland to dust bowls and forests to ashes. Picture last fall's California wildfires as a regular thing. Or imagine similar disasters destabilizing nuclear powers such as Pakistan or Russia—it's easy to see why the Pentagon has become interested in abrupt climate change."

It is unfortunate that George Bush Jr. failed to listen to the advice of the wisest of Pentagon leaders, and instead opposed efforts to tackle climate change. Now we are in even worse shape after four years of accelerated emissions. What will it take to make the needed change in the emissions caused by the polluting practices of industrialized humanity?

Chapter 34
Feedback Loops

The Pentagon report warned of positive feedback loops, whose scale is prompting such dire warning of climate catastrophe. As the Earth warms with increased greenhouse gas emissions, it is triggering various climate-influencing systems, also called feedbacks, to change from being cooling mechanisms into warming mechanisms. This may be a sign that the planet is lurching into an entirely different climate, one far warmer than the planet we are on today.

Negative feedbacks are systems that reverse warming or cooling trends, and tend to stabilize the climate. The increased uptake of carbon by oceans and forests in response to our increasing their levels in the atmosphere is a negative feedback loop, and has helped to reduce the amount of warming we otherwise would have seen thus far. Positive feedbacks are mechanisms in which the warming triggers more warming.

Nearly every major climatic feedback system, of which there are many, is now acting as a positive feedback, meaning that the warming we are causing in adding greenhouse gas emissions is triggering more warming. Most of the negative feedback mechanisms are declining or even reversing into becoming positive feedbacks. The North Atlantic Ocean absorbed half as much Carbon Dioxide as it did in the mid 1990s (Reuters, April 17, 2008). As the ocean gets increasingly saturated, its ability to sequester carbon goes down. Meanwhile, the carbon forms carbonic acid that threatens mass extinction in the ocean.

The albedo effect is the ratio of light that is reflected as the sun hits the planet. With a higher albedo, more sunlight is reflected, cooling the surface. Fresh snow has an albedo of about 90 percent, whereas charcoal has 4 percent.

The white polar ice and snowy Arctic tundra reflects most of the sun's energy back to space, while open water and vegetation absorb most of that energy. The melting of the ice through global warming is creating a positive feedback loop, causing more warming. Scientists warn that the melting of Arctic ice threatens to cause as much warming as that caused by all of our emissions thus far. And it is only one of numerous positive feedback effects that threaten to worsen exponentially as the earth warms at an accelerating pace.

Climatologist Dr. Josephino Comiso said "this is a major effect in the Arctic region. If you look at the different regions, that is why the Arctic and Antarctic are so different. In the Antarctic, you have a big continent, greater than the size of the United States, and that's always covered by snow, so the continent is high. In the summer, when the sea ice retreats, the sea ice cover is reduced to about 25 percent of the winter ice cover. The impact of the sea ice retreat is not quite the impact as in the Arctic, because in the Arctic, what you have as a reflector of solar radiation is the sea ice cover. Greenland is quite small compared to Antarctica. Just sea ice cover and

Greenland together are the only entities in the Arctic that reflect solar energy and keeps a lot of heat from being absorbed by the Earth.

"Now as the ice retreats, and what we have observed recently is that the summer sea ice is retreating quite rapidly at the rate of 9 to 10 percent per decade. As that ice retreats, you have less and less of that reflector in the Arctic. During the peak of summer for example when you have 24-hours a day of solar radiation, as you get less and less ice there, then you get more and more solar radiation to the Arctic Ocean. As the Arctic Ocean gets warmer, then the impact there is the freeze up of sea ice is later and later, and the melt-off of sea ice comes sooner as well. You have longer periods of melt, and therefore you have thinner ice. Which then becomes even more vulnerable to melting, so this is what we have been calling feedback.

"There is certainly something going on in the region. I think that there has to be a concern. Especially because there are these various feedbacks that I have been talking about. Like if you look in the summer if you look at the perennial ice, and all of the radiation from the sun, and you see the large fraction that would be absorbed by the ocean."

"What happens is when you have less ice you will have more exposed ocean. Then you're replacing high-albedo ice with low-albedo ocean, so you melt your ice early in these shallow seas in the Arctic, and you have sun shining on them that could warm up that water quite a bit. So then what you expect is in the fall when the ice starts to grow again it just takes a lot longer because now it has to cool down that ocean that has warmed up quite a bit in the summertime because it was exposed to the sun and less was reflected back by the ice cover.

"The ocean absorbs more sunlight. The ocean ends up warming up because the seas are fairly shallow all around the Arctic edge. So each year this continues to happen you may get a point where there is no sea ice.

Like the retreating pack ice, changing open tundra to forest greatly increases the amount of energy that the Earth absorbs, said Alaska Conservation Solutions President Deborah Williams. "The vegetation does change the albedo. Alaska vegetation in particular, like the black spruce, is well adapted to absorb heat. It is very dark and absorbs a lot of heat. When you push black spruce and other vegetation north, you have this positive feedback where it accelerates the impacts of global warming. So it is quite extraordinary."

Another major feedback loop is the massive amount of carbon and methane in the melting permafrost, that is beginning to be released. The permafrost, or year-round frozen ground, in the lands surrounding the Arctic Ocean, holds an estimated 1,500 gigatons of carbon (Commonwealth Scientific and Industrial Research Organization, 2008). A gigaton is a billion tons. Greenhouse gas emissions from human activities release about 25 gigatons of carbon annually. Frozen, the carbon in the permafrost is trapped, like coal, oil and natural gas are until they are burned.

Temperature increases of 8 degrees Fahrenheit in the Arctic lands in the past two decades is causing a rapid and accelerating melt of the permafrost, releasing that carbon and methane.

University of Alaska temperature measurements 66 feet deep in the permafrost rose nearly four tenths of a degree from 2006 to 2007. That much change, that deep, in one year, is very alarming.

The 2007 IPCC findings, which are already quite startling, did not take this additional carbon into account. In Siberia, warming is creating shadow lakes. With a higher temperature because of the dark water surface, permafrost around these lakes melts and collapses into the lakes, releasing methane. University of Alaska at Fairbanks Biologist Katey Walter warned that the release of methane from the melting permafrost is a "ticking time bomb."

Similar mass methane releases may have been responsible for the Permian-Triasic extinction event 251.4 million years ago. Earth's most severe extinction event, up to 96 percent of all marine species and 70 percent of land vertebrate became extinct. The same effect may also have caused the Paleocene-Eocene Thermal Maximum extinction event 55 million years ago. "A catastrophe lays buried under the permafrost," said National Snow and Ice Data Center scientist Ted Scanbos.

"A global tragedy of monumental proportions is unfolding at the top of the world, and the human race is all but oblivious to what's happening," wrote Author Jeremy Rifkin (Los Angeles Times, 2007).

The warnings have been given, but because of the entrenched, powerful fossil fuel industry cause of the problem, and their amazing power of disinformation and political influence, our emissions continue the rise.

Chapter 35

Methane Clathrate Mega-Threat

As if the methane in the permafrost and the changing albedo threats were not enough, a far greater positive feedback is beginning to be unleashed: frozen methane hydrates on the ocean floor.

About 3,000 times as much methane as is currently in the atmosphere is locked up on the ocean floor as frozen methane clathrates, also called hydrates. Kept frozen by pressure and cold, the few degrees temperature rise we are now experiencing is beginning to release, them. The potential of this is horrific.

The scale is far greater than both the permafrost melting and albedo effects combined. When the methane is released slowly in small bubbles from the ocean floor, it is consumed by microorganisms and does not reach the atmosphere.

But when it bursts from the ocean floor in large amounts, it bubbles to the surface, and into the atmosphere. This very event was documented by scientists off the coast of Santa Barbara, California in 2002. A massive released of a 5,000 cubic feet was videotaped with green die inserted. It rose 60 feet to the surface in seven seconds. The scientists reported that it sounded like a freight train.

Many scientists believe methane hydrates caused the excessive warmth 55 million years ago that wiped out most species. The massive amount of methane that we are potentially triggering the release of -- trillions of tons -- should prompt a massive focus to stop burning fossil fuels.

Just a few degrees in warmth can cause their release. Many clathrates lie in sub-Arctic waters, whose temperatures are rising several degrees. With several more degrees of warming predicted in coming years, we are in dangerous times. When melted, one cubic meter of clathrate turns into 170 cubic meters of natural gas (methane).

In addition to the greenhouse gas emission impact of the methane, they can also cause a physical disturbance as they burst to the surface. Lightning or any ignition source, such as a boat engine could ignite the methane, causing a massive explosion that scientists say could exceed the size of a nuclear bomb.

Methane hydrate melting appears likely to have been a factor in the Permian-Triassic mass extinction 250 million years ago when 95 percent of the species on the planet were lost forever, as well as the extinction 55 million years ago.

Methane hydrates are known to be instable by nature, and easy to disturb with physical impacts like oil and gas exploration and development, or by rising temperatures. Much geological evidence shows that rising methane levels comes just behind rising temperatures.

Some have suggested mining the methane clathrates, which contain twice as much energy as all of the known fossil fuel reserves. But scientists warn this could trigger a runaway greenhouse effect. As we should think of both methane clathrates and fossil fuels as carbon reserves, we should not burn either. That, of course, will take some major changes as we are burning them currently at an increasing pace. That is where we must collectively sound the alarm about the urgency of this issue.

As horrific as the methane hydrate threat is with just a few degrees more global warmth, there are still many more positive feedbacks expected from the dying and burning tropical rainforest, from the boreal and temperate forests that are dying and burning and much more.

Past geological evidence is clear that when our planet starts warming, it tends to warm at an extraordinarily fast, accelerating pace until a new equilibrium is reached, one our species has never experienced. The pace at which we are increasing greenhouse gases vastly exceeds anything in our planet's past, including the most extreme events ever seen that have caused massive extinctions. We must make this the number one issue, and put the brakes on the runaway climate crisis while we still have a chance.

Chapter 36

Climate Change Interferes With Oil Exploration

Back in Kaktovik, I bid farewell to Stacey, Robert and Waldo Arms' staff, and hiked over to the airstrip. I boarded an 8-seat, double-propped plane. We flew along the coast. For the first sixty miles, all was wilderness: a patchwork of tundra wetlands and lakes, and the shore of the Arctic Ocean. The coastal plain is an oasis of life: an amazing patchwork of wetlands, streams and wide rivers braided into dozens of channels.

As far as I could see to the north was open water: hundreds of miles of open water where just a few decades ago the ice rarely receded more than 15 miles from shore at the most.

The scale of the change just in the 14 years between my 1991 journey and 2005 was amazing. This was no random occurrence but the trend toward global disaster that we should have heeded decades ago.

Now, the change is so fast and has gone so far that we are doomed to major world-wide damage. But we cannot give up. Instead, we must motivate to move faster to stop the burning of fossil fuels, the destruction of old growth forests and do something about livestock and chemical agriculture emissions. Because if we do not act fast, major worldwide damage will turn into the likely obliteration of life as we know it: leaving a desert wasteland to future generations.

For miles from Kaktovik, the coastal plain was as perfect as all time: an extraordinarily rare expanse of wilderness, with not a single human structure in sight. The Hulahula River snaked across the open tundra, separating into more than a dozen channels of blue water over wide gravel beds, with green tundra in between.

After the river, the plain was full of lakes of various sizes, along with wetlands. There was about as much lake cover as there was tundra. Smaller ponds looked like beads connected with streams that opened into a wider channel before entering large lakes. Smaller streams snaked across the tundra. I could see far up into their headwaters, all complete wilderness in need of permanent protection.

As we passed over the Canning River from the Arctic National Wildlife Refuge into state-owned lands that have been leased to the oil companies, the first scars stood out like a compound fracture. Mile after mile of roads, oil rigs and pipelines extended in all directions: even into the ocean.

Though the oil and natural gas was not visible, I knew millions of barrels were pumped every week from the North Slope of Alaska. It is ironic that one of the main culprits of the climate crisis – the oil and gas industry – is now finding its winter exploration hindered by the very warming they are in large part causing.

A well documented impact of climate change is the shortening of the season in which the oil industry is allowed to explore and develop for oil. In order to protect the fragile tundra, federal and state regulators required the oil industry to only develop when the ground was frozen 12 inches deep and had 6 inches of snow cover.

While in 1970 more than 200 days met those standards, with climate change, only 103 days met the criteria in 2002. Cox News Service reported in 2003, "global warming, which most climate experts blame mainly on large-scale burning of fossil fuels, is interfering with efforts in Alaska to discover yet more oil."

Instead of realizing the significance of this change as a global need to reduce greenhouse gases, and invest in the proven solar and wind alternative rather than more fossil fuels, the state of Alaska has instead changed the rules, to allow more damage to the tundra by further opening the season of oil devastation to the fragile north slope by up to 6 weeks.

Initially, the study of tundra impacts was to "refine our understanding of the tundra's disturbances." But when the study was released in December, 2004, the state said it needed to change the standards to open the season longer because it is "a problem for Alaskans who rely on the oil industry: in other words, for all of us. We need to lengthen the season." The December 3, 2004 press release by the state of Alaska admitted that the change in seasons was caused by "warming weather."

Environmental groups charge that the existing standards were not protecting the tundra, and the changes will only increase already unacceptable levels of damage. One of the lead environmentalists working on this issue happens to be a college friend of mine. Kelly

Hill-Scanlon and I were in the Ohio University Ecology Club, which succeeded in getting recycling institutionalized at our university and protecting lands around Athens back in the late 80s and early 90s. Now she is the Arctic Campaigner for the Northern Alaska Environmental Center in Fairbanks, Alaska. She is also monitoring the test sites that the state is using to make new standards for oil development.

I stayed with Hill-Scanlon, her husband, a fisheries biologist with the state, and their three-month-old daughter during my visit to Fairbanks in late August, 2004. Hill-Scanlon explained why the state of Alaska is so predisposed to wanting oil development. “The state of Alaska in the constitution says lands for the benefit of the citizens of the state—that’s meant mine it, drill it, log it. The state of Alaska owns all subsurface rights even for federal lands. That is why the state is so money hungry for the drilling of federal lands.

“That’s something we’re trying to combat. The benefit does not have to mean economic gain. The caribou are a benefit. You can’t put a dollar value on it. The so-called economic gain is all these foreign companies coming into our state and taking our resources. Is that really benefiting our state? It’s disrupting the caribou in the western Arctic.

“The work season used to go from November through early April. Over the last 5 years, that window has declined because of global warming to where sometimes January is the first date to travel. That does not work for the oil companies. They need to build ice roads first. The Department of Energy gave a grant to do a study by the Department of Natural Resources. In the proposal it says 'due to global warming. The oil companies agree that global warming is real because they have signed on.

"They have a plot 3 miles outside of Deadhorse and one in Happy Valley." Kelly witnessed the plots. "They have 30 plots of land 50 meters long. In July, 2003, the DNR got baseline data from the plots—the vegetation, the species, the soil, the moisture level, heights of plants.

"There were testing dates in September, October, November, December and January. They tested 5 vehicles and left some untouched. They would do a figure 8. That was this past winter (2003-2004). DNR went back up this summer and re-measured everything.

"A bulldozer in October ripped up the ground. It's not only a year since the damage. Tundra damage often does not show for more than a year. One problem is there is just one year of study. They are trying to get Yale to pick it up. The governor wanted the results released in September. It is basically a rush job. DNR said not by September but by November. Scientifically there is a problem. There is no peer review.

"They're going to do peer review, after it is released. That's like letting the fox in the henhouse and then trying to get it out. We know the political motivation is to open the tundra earlier. The scientists can only extrapolate it to those plots, not to all the areas of the north slope.

I went there last summer. From the October tests, there was significant damage. The bulldozer track was like mud. We want to make sure it is sound science, peer reviewed and we want to make sure it will really protect the tundra and not just buy a few more days for the oil companies.

Forbes found that the tundra will continue to show damage from vehicle disturbance for two years after the impact. Thus, by only studying the impacts a few months after the vehicles drive through, they are missing the long-term effects. Hill-Scanlon said that they failed to employ the proper experts. Forbes concurred, with questions as to the expertise of the scientists, and a volley of criticisms with a study that has far more to do with the interests of the oil industry and very little science.

Alaska Conservation Solutions President Deborah Williams also noted impacts on oil and gas development. "The other infrastructure story is how climate change is reducing the period in which oil companies can develop oil and gas using ice on the North Slope. The number of days has been reduced drastically. The state of Alaska documented this well. So for years and years the oil industry has relied on ice roads, ice pads and ice technology to do relatively low cost, low impact exploration on the North Slope. Now the season is becoming so short, that it is no longer becoming economic in some instances for the oil industry to do ice-based oil exploration and development. So they are looking at other technologies with rigor.

The state of Alaska is trying to determine how much ice and snow is necessary to do oil development. They are trying to redefine that down so they have a longer season. That is a bit of an ironic impact of global warming because the primary cause of global warming, oil and gas development is now being impacted by it, and now the industry is trying to come up with exceptions to deal with that."

Wohlforth said, "They are the ones producing the fossil fuels, and the impacts of the fossil fuels appears to be making the ground thaw so that they cannot do their job."

The oil and gas industry already cannot do its activities without damaging the tundra, so making it less stringent will only further increase the damage that is already at unacceptable levels, said Pam Miller of Arctic Connections. She warned of more alarming impacts as toxic waste dumps that were frozen into the permafrost, and considered contained, are now melting and being released into the surrounding environment.

"What they find is around the drill well, where it goes down into the ground, the permafrost is melted. And you typically have a pretty contaminated site there. They consider in the Prudhoe field every one of those is a contaminated site. So when you get to the point when ultimately the oil runs out and they are no longer connected and they are not going to pump any more from the ground, you are going to have lots of those sorts of sites.

"Another thing, there are situations where in the past the military buried a lot of toxic and radioactive waste in the Arctic. In the National Petroleum Reserve Area (NPRA) in northwest Alaska there are a lot of old waste dumps and there are also military sites where they had landfills off Kaktovik in Barter Island and they are eroding into the ocean now. The sites that may have been 'sealed into the permafrost,' are now melting and leaching out. There are a couple places where they've closed out a lot of the drilling waste pits and dug a huge big hole and thought they could secure that material in a site that would be frozen, and it actually didn't work like that. And to presume that a site that now may be frozen into the ground, that is not a safe assumption for a way out.

"The oil industries own operations are now coming back to bite them. Because of the global warming, their ability to rely on ice roads to access either development sites that might be built without a road -- there are a couple of those already -- or for their new exploration programs, they are realizing that they don't have the working season that they used to. But what they are wanting to do is very political, in my opinion.

"On the state land with the state's priority of land use, they might be willing to trash a lot more tundra. And I quite frankly think that they'll figure out a way no matter what the results say to get out there earlier in the state lands. Clearly industry wants its way, and is getting it politically in the NPRA and on the state lands. It is not going to solve the problem that there is not a long season for them to build ice roads.

"With the existing standards, there is still damage from off road vehicles, seismic exploration, cat trains and fort camps that they haul around. From the repeated laying down of ice roads, from repeated use of land from the rollagon trails, there are scars across the state land even with that state standard.

Guntner Weller said climate change is impacting infrastructure, roads and pipelines. "There are huge problems when the permafrost melts, and industrial facilities collapse. For example with the Trans Alaskan Pipeline, we've talked to the engineers. About half of the pipeline is buried, the other half is above ground supported by a steel frame. The steel frame is artificially kept frozen through a refrigerant, so the base of the steel structures are embedded in permafrost. As it gets warmer, the permafrost table goes down, the support begins to sink and they have to be re-engineered, and drilled deeper. These are impacts that are not insurmountable, but they cost more money. There is an economic impact."

The National Academy of Sciences 2002 Cumulative Environmental Impact study of human effects on the north slope of Alaska reports that, "the climate is expected to warm so rapidly that the effects of current activities could be much greater on the permafrost landscape than would be the case if the climate were relatively stable." The academy of sciences report continues, "The largest human contribution to climate warming is the burning of fossil fuel hydrocarbon fuels. …The resultant climate change affects the North Slope probably more than lower latitude areas…. It is an important factor that must be considered."

Miller said, "There has never been enough water to build ice roads in the Arctic Refuge to begin with. The fact that ice roads aren't feasible from the industry perspective because the season is so short, it could exacerbate the limitations there certainly, but there was already a problem. There is already a problem with Alaskan snow cover. You can find a six-inch average in the Arctic Refuge most winters, but it didn't protect the tundra when they implemented that standard."

As far as a solution to climate change, Miller asked, "are we going to take a far-sighted view to change towards a new future sustainable energy policy that can reduce our fossil fuel emissions that are contributing to global warming? And from that standpoint, it doesn't make sense to sacrifice a unique resource like the Arctic Refuge, when our broader policy needs to make the shift from the oil age of fossil fuel development to what is needed now." Miller said the investment that is needed now is in renewables and conservation to get clean energy faster with the goal of reducing fossil fuel emissions.

Chapter 37

Prudhoe Bay: witness to the source of climate Armageddon

Drill pads, small rectangular buildings, roads, parking lots, rusty machinery, cars, trucks, pipelines, toxic ponds and pits, waste dumps, air strips, hangars and warehouses made for an ugly, industrial sight all the way from the Canning River to Prudhoe Bay. After we landed I strapped together my mammoth pack, and began searching for a camping spot. I found a small strip of level ground, and set up my tent. Then I walked to find out where to catch the bus tour of the drill field.

Walking around Deadhorse from my camp, the landscape was quite ugly, with vehicles and buildings of clashing colors. Noisy, polluting vehicles drove by far to close for comfort, with narrow roads dropping into ditches filled with oily “water.” Engines roared in all directions. This is not a place I would visit for pleasure.

The bus tour operator was a British Petroleum Security Officer. He gave us a presentation, talking about how he works two weeks on, then gets two weeks off. The oil companies provide his air transport from Anchorage, his food and his room and board. They also provide complete brainwashing: turning the security guard into a power hungry drill sergeant who worships every move of the pollution industry for which he works.

I initially sat on the left side of the bus. But the window was completely covered with black soot. I stayed in that seat to video a vast drill field with massive flames of burning gas. What a waste of energy. We are burning up hundreds of millions of years of stored carbon, and somehow think it is alright. Even if it were not for the climate catastrophe that we are causing, the fact that it took about 300 million years to make most of the oil that we are burning in a matter of decades is incredibly selfish and greedy.

Spotting a cleaner window by a vacant seat, I unbuckled my seat belt and scooted across to the other side. The bus driver said "sir I am going to have to ask you to stay in your seat. You need to stay right there in your seat. We do have rules and regulations here, and one of them is to stay in your seat with your seatbelt on."

"I'm sorry, it's a cleaner window over here," I said meekly.

"I am going to have to ask that if you cannot abide by those rules we will not go on to the field."

The driver said that the power stations in Prudhoe Bay provide an equivalent amount of power as is used by Fairbanks, the second largest city in the state. Imagine how many wind generators and solar panels could have been made with the tens of billions of dollars invested in getting oil from the North Slope of Alaska.

The so-called directional drilling improvements that theoretically will constrain the sprawl are far more propaganda than fact: they have only decreased the size of drill pads by 10 percent in 30 years, while oil spills are increasing. More than 550 oil spills occur every year around Prudhoe Bay: more than one a day. The consequences are now becoming clear. In 2002, the National

Academy of Sciences found that drilling around Prudhoe Bay had a negative impact on the caribou, bowhead whale and almost every other wildlife species.

I've seen this devastation first hand when I backpacked hundreds of miles through the oil fields of Prudhoe Bay, and into the Arctic National Wildlife Refuge in 1991. Prudhoe Bay is an industrial nightmare with massive development and toxic waste spread over more than a thousand square miles.

They have dredged enough gravel to fill 90,000 football fields to a height of 3 feet for all of this development. Dredging for gravel is done in rivers and streams and devastates their ecology. There are gravel roads and pipelines all over, gigantic pumping stations, a massive network of roads, ports, airports and toxic pits.

Existing oil development on the Alaskan North Slope has more than 4,800 exploratory and production wells, 223 production and exploratory drill pads, more than 500 miles of roads, 1,800 miles of pipes over 600 miles of pipeline corridors, two refineries, 20 airports, 107 gravel pads for living quarters and other support facilities, five docks and gravel causeways, 36 gravel mines, 28 production plants, gas processing facilities, seawater treatment plants and power plants.

Oil companies caused more than 4,532 spills between 1996 and 2004 totaling more than 1.9 million gallons of toxic substances. The most commonly spilled pollutants are diesel, crude and hydraulic oil. More than forty different toxic wastes have been spilled across a thousand square miles in all directions from Prudhoe Bay. The number of spills has been dramatically increasing, from an average of 400 spills per year in 2000 to an average of more than 550 spills per year in 2004.

Air pollution around Prudhoe Bay exceeds that of Washington, DC, and nitrogen dioxide levels are twice that of DC. The oil industry emits 70,000 tons of nitrogen oxides, which cause acid rain and smog, 6,199 tons of particulate matter, 1,470 tons of sulfur dioxide, 11,560 tons of carbon monoxide, 2,647 tons of volatile organic compounds, 24,000 metric tons of methane and up to 40 million metric tons of carbon dioxide annually.

Chapter 38

Lurching into Inhospitable Greenhouse Earth

Seeing the scale of the oil exploitation as I took a bus tour around Prudhoe Bay, along with my recent witness of mountaintop removal coal mining sites, the fact that this is massively changing our climate is not only plausible, but logical. Mile after mile we drove past massive gravel pads, drill fields the size of shopping centers, warehouses, trucks by the hundreds, pipelines stretching into the horizon and roads, always roads, with gas guzzlers kicking up clouds of toxic dust.

Geologists have found that the Earth has two relatively stable states, one with polar ice, termed an Icehouse Earth, and the other, a Greenhouse Earth, where the polar caps are melted. Through the mass burning of fossil fuels and the destruction of old-growth forests, we may have flipped the switch, or be teetering on the brink of doing so, whereby nature takes over in pushing us toward temperatures that the Earth has not seen for 55 million years.

The Icehouse Earth state includes two sub-states, glacial times and interglacial time periods, as we currently are in – or were in until we began the mass release of greenhouse gases. The jumps between the two main climate states have always been rapid, and combined with mass extinction.

Imagine the scale of the change. British Scientists Sir James Lovelock predicts that this change will turn the equatorial regions of the world into uninhabitable desert landscapes similar to lifeless Mars. He predicts the desertification will expand into southern Europe, North America and Australia, with a massive death toll. "We will be lucky if 20 percent of us survive what is coming," Lovelock said. "We should be scared stiff."

Ice cores reveal that when it starts to warm, it tends to warm very fast. This has happened four times in the last 400,000 years, as the planet has gone from ice age to interglacial time periods. Being in an interglacial, the greenhouse gases that we have added to the atmosphere appear to have triggered multiple mechanisms that are beginning to launch the planet into a new, much warmer system: one not seen for 55 million years.

"Long term history of the Earth suggests the existence of hot and cold stable states," Lovelock said. "What the geologists refer to as the greenhouses and the ice houses." These are also called attractor states. Dr. Alder Fuller, founder of Euglena Academy, likened attractor states to a brick. When you tip the brick, it will either fall back into place, or if you pass the tipping point, it will tip over. In the case of the climate, this new position is a new climatic system, one vastly hotter.

Like the Pentagon warned, it is like the tipping of a canoe. Once it gets going, gravity takes over, flipping the boat upside down. As positive feedback loops are kicking in, we may be moving toward the next stable state, when temperatures are twice as high as the difference between the last ice age and now. More and more evidence is building that not only is this a possible risk: it appears to be exactly what is happening. British Scientist James Lovelock is convinced that we are already inevitably going to launch into this other state.

With his push for nuclear energy, many environmentalists are reluctant to listen to Lovelock, and for good reason. Nuclear is a wrongheaded approach to energy, and nuclear waste is extraordinarily dangerous. Uranium is a non-renewable resource requiring enormous amounts of fossil fuels to mine and refine. And, nuclear waste cannot be safely contained for millions of years. Solar and wind energy are a proven solution to our energy needs, and both are cheaper than nuclear.

Still, it is important to listen to Lovelock's message on climate change. At the Clearcutting the Climate Conference in Eugene, Oregon, sponsored by the Native Forest Council, Dr. Alder Fuller said, "If you don't like Lovelock's solutions, just ignore them. But take his assessment of the crisis to heart. He knows more about the climate system than anyone I am convinced of it."

"We have probably passed a critical threshold (tipping point) for climate change that will soon transform Earth into a different planet with a more violent & chaotic climate that hasn't existed for 55 million years," Fuller said. "We must begin planning to adapt. The speed could be one or two decades for major changes. The temperature increase could be twice that from the last ice age to the present."

Average interglacial methane levels were 650 ppb. We are now at 1780 ppb, two and a half times higher. Carbon dioxide levels are far higher than at any time in the last 800,000 years. The highest that CO2 levels have been during that time is 300 ppm. We are now at 386 and increasing 2.6 ppm per year at an accelerating rate.

There is a 50-year lag time between atmospheric gas stabilization and temperature stabilization, because water absorbs heat 20 times more than air. It takes a lot longer for the oceans to warm up to an average temperature range. The fastest we have warmed previously is 30 ppm CO2 per thousand years. We now increase it at the rate of 30 ppm every 12 years. Fuller warned of "Rapid, violent and chaotic" climate changes with "disruption of agriculture, water availability (and) ecosystems."

Lovelock said that the last time carbon dioxide went up one to two terratons, because of a geological accident, temperatures rose 8 degrees Celsius and 5 degrees at the equator, 55 million years ago. This change was done over as much as 10,000 years. It took about 200,000 years for conditions to return to their previous state.

We have added about half a terraton during the 20th century, that Lovelock says has already launched us on the path toward the new, hotter stable state. The scientist warns that we and the Earth (through positive feedback mechanisms) are likely to release another terraton of carbon shortly. This would cause the equivalent change as occurred 55 million years ago, only over a period of 200 years, instead of 10,000 years. Doing it so quickly is what is acidifying the oceans, by creating carbonic acid with the increased carbon dioxide. When it happened 55 million years ago, the ocean had time to mix, and the acid was neutralized by alkalinity in deep waters.

"We are in the strange position of living in a planet… where the change is happening too rapid for us to react to it," Lovelock said. The positive feedback of the Arctic region is nearly equivalent to all of the pollution we have already emitted, he said.

Researchers have found nine "tipping elements" that could rapidly cause major shifts in Earth's climate as they are triggered by the increased temperatures caused by greenhouse gas emissions. The proceedings by more than 50 scientists were published in the National Academy of Sciences journal in early 2008. They lend weight to Lovelock's warning that we are now lurching toward the next stable state climate system vastly warmer than today.

"In some regions anthropogenic forcing on the climate systems could kick start abrupt and potentially irreversible changes" (Potsdam Institute for Climate Impact Research, February 4, 2008). "For these sub-systems of the Earth system the researchers introduce the term 'tipping element.'"

The tipping elements that they found, and the time it will take to undergo major transition are:

- Melting of Arctic sea-ice (about 10 years)
- Decay of the Greenland ice sheet (about 300 years)
- Collapse of the West Antarctic ice sheet (about 300 years)

- Collapse of the Atlantic thermohaline circulation (about 100 years)
- Increase in the El Nino Southern Oscillation (about 100 years)
- Collapse of the Indian summer monsoon
- Disruption of the West African monsoon (about 10 years)
- Dieback of the Amazon rainforest (about 50 years)
- Dieback of the Boreal Forest (about 50 years)

The research team warns that people should not be lulled into a false security because of the seemingly slow warming that has been seen in past decades. It is not slow on an evolutionary time scale, but extraordinarily fast. And we appear to have set in motion a massive change that we will not likely be able to stop. Still, we must mobilize rapidly to halt the burning of fossil fuels and the destruction of old growth forests, as fast as possible.

But because of the massive scale of the damages predicted, scientists suggest that we also work on mitigating the effects of the change. While we might be able to survive heat waves in air conditioning indoors (at least those who can afford it), the food that we need to survive may be far harder to grow, and water to drink could run out.

Many studies have found that climate models underestimate the actual warming. What usually happens is that the climate is very complex, and influenced by so many factors that scientists cannot include all of them in their models. Almost all of the influences are positive feedbacks, pushing the climate into a new, inhospitably hot state. So when factors are missed, they miss the true extent of the warming potential.

In February, 2008, scientists announced evidence that prehistoric tropical Atlantic Ocean temperatures may have reached 107 degrees Fahrenheit (42 degrees Celsius), 25 degrees Fahrenheit warmer than today. These temperatures appear to have occurred between 84 and 100 million years ago, when carbon dioxide levels were between 1,300 and 2,300 parts per million.

Models do not find temperatures that high when 1,300 or 2,300 parts per million carbon dioxide levels are plugged in, said Woods Hole Oceanographic Institution paleoclimatologist Karen Bice. "Current models may be missing a critical factor that amplifies heating," Bice said. Methane released from oceans and wetlands may have been a factor. Paleooceanographer Mark Leckie, who reviewed the study, said "Karen's research should be another wake-up call to the rate at which we are changing the system today."

Chapter 39

Dire Future

With climate change expected to increase as more greenhouse gases are emitted each year, the future of the Arctic and Alaska looks bleak, with more dying forests, shrinking glaciers, retreating pack ice and crumbling shorelines. And that is just with the climate changing at the rate we have seen thus far. What scientists are all finding, however, is that the rate of warming is increasing exponentially each decade, as feedback loops add massive warming in addition to the increased greenhouse gas emissions.

Williams warned that many of the effects of climate change cannot be mitigated. "Some of (climate change's impacts) can be mitigated. Oh well, you can move villages. You can rebuild highways. There is certain mitigation that you can do. But other things can't be mitigated. If polar bears become extinct because of global warming, if native cultures become extinct because of global warming, these impacts can't be mitigated. And I think that is a very important point to make. Because global warming is more palatable if you think, 'Oh we'll just throw enough money in mitigation.' We'll just deal with it as the impacts become known.

"But, if we go into global warming with the knowledge that there is a likelihood, if not the inevitability, of profound impacts that cannot be mitigated, I think it changes the way we look at our responsibility to address global warming now. The other question with mitigation too, is if we talk about moving villages at $100 million a pop here, $100 million a pop there, and all of the other costs associated with global warming, the impacts on fisheries, the impacts on Agriculture, wildlife and disease, and start adding up those costs, pretty soon I think that we as a society can conclude that the costs of prevention are a lot less than the costs of global warming. I really do believe that Alaska helps demonstrate both the extent of the costs that you can mitigate, but also the extent of the costs you cannot mitigate. Alaska's greatest lesson -- global warming is profoundly costly."

Fran Mauer called the impacts of climate change in the Arctic "unprecedented in the history of the planet."

Forest Ecologist Glenn Juday warned that all the predictions of climate change's future impacts would lead to a massive loss of forests in Alaska, far more than has already been seen.

"All of the scenarios basically raise the temperatures to levels where we would begin to see by the end of the century the temperature threshold at which the trees would not be able to grow on certain sites. So here is what I project. We would see continuing levels of very large-scale outbreaks of insects that kill trees.

"That has always happened, there has always been some of that around. But we are going to see a larger percent of our landscape is going to be made up of big patches of dead trees under conditions like the scenarios produced.

"Number two, there would very likely then be a lot of fires, because those sites would probably burn. And the forests, once they got going would be burning lower moisture content fuels so they would burn more sites and burn more thoroughly. So basically we would have larger and more frequent forest disturbance from insects and from fires. So we would have fewer older forests in any given time and more younger forest. And as we got near the end of the scenario time we would see forests unable to regenerate in certain landscape types so natural grasslands would become a more important part of the landscape." He predicted large prairies where once there were forests.

"You would see permafrost thawing. Fifty to fifty-five percent of the boreal forest is made up of black spruce forest dominated stands. Almost all of that is under-laid with permafrost. You would start to see a substantial amount of that permafrost start to thaw. You would see big sink holes and low spots that would begin to collect water around the trees and you would see a lot of that in these low elevation black spruce permafrost types, and then eventually some of it would drain out, some of it would remain ponded wetlands, permanent wetlands. You would generally see a drying out of the landscape and then you would see as well forests moving up in elevation and possibly getting north of the Brooks Range barrier and getting to establish on favorable sites in the North Slope tundra.

"The scenarios produced climates that would allow white spruce to grow in the tundra foothills region of the North Slope." The chance of seeds getting there would be unpredictable, he said, so the actual trees may take longer to appear, but they would be able to grow there under the new climate models.

"Researchers in Siberia have recovered some frozen tree remains from the permafrost," Juday continued. "From 8,000 to 6,000 years ago, there were trees all the way to the coast of the Arctic Ocean in Siberia. The summers were warmer during that period of time, and over a period of thousands of years, the trees gradually retreated. It happened in major cooling episodes.

"If the scenarios of warming are correct, it is very reasonable to project that trees will make it all the way to the Arctic coast, but it probably would not be the full canopy. It would likely be a forest tundra, with trees in more favorable sites, interspersed with tundra in between. The tundra would likely change with fewer tussocks. The shrubs would get a lot taller. It would produce some things helpful to caribou survival, some things negative, but most tundra animals would see a squeeze on their populations."

"It is a pretty straightforward proposition. You don't have to believe in a doubling of earth's population," Juday said. "If we humans continue to pump out greenhouse gas emissions, if we just continue what we are doing now, we will build up a significant problem for the survival of elements of biodiversity. We either come to some limitation or slow the rates or reverse the increases or we will have to deal with those things.

"The only debate really is how long. Now, could humans survive on an earth with a severely greenhouse effect-impacted climate, well probably yes, but there would be major effects.

"We have to recognize that a significant advance of the forest into the tundra would amplify the climate warming. With spruce forests, 80-90 percent of the sun's energy is absorbed, whereas snow-covered tundra reflects 80-90 percent. Converting smooth, dry snow-covered tundra to rough, dark spruce covered forest would produce more warming."

Compounding the problem more, Juday said "the amount of carbon stored in the boreal forests and forest soils is very large. The amount that could be converted to methane or CO2 is a substantial fraction of what is already in the atmosphere. This would amplify the warming. So it looks like already one of the things that land managers in boreal regions need to look at is ways to sequester carbon dioxide."

Climatologist Dr. Josefino Comiso also warned of accelerating impacts of climate change. "Modeling studies confirm that if you have so much increase in carbon dioxide in the atmosphere, you have an increase of temperature as well on account of the greenhouse effect and also the feedbacks. You get the ice retreating for example, you get more solar energy to the surface of the Earth and this furthers warming."

"We really have to be serious about this impact. Because even though we are sure now about the large variability in the perennial ice cover in the Arctic, the trend is very significant toward a reduction of ice. When we deal with the planet, we need to be very careful with the activities of human beings, and make sure that we are not the cause of a permanent harm on this planet, which is the only planet that we know of that has intelligent life in the universe. We don't want to take the risk of losing that because we were negligent for doing what is right for our system."

Juday said, "The message to other folks in other parts of the world who are at risk is get ready. Because when it happens, you've got to adjust. You'd better get ready. I think you've seen in the last several years in the West extensive fires have occurred more frequently over a greater area with a greater severity and people had been planning on using those trees for various things. They put their homes in those various places. And nature hasn't been too kind to those expectations that you had. So be prepared."

What makes the situation so dire is the entrenched commitment of anti-environmental lobbying and policies of coal, oil, auto and airline industries and lawmakers (mostly Republican) that have blocked efforts like the Kyoto Protocol that would begin to make some real cuts in global warming gas emissions. Syndicate columnist Molly Ivins wrote, "The response of Alaska's Republican right (to climate change) is to slash and burn, subsidize mining and give tax breaks to oil and gas companies for exploration. Cronyism and favors for special-interest groups have become the hallmarks of state government. Gov. Frank Murkowski even appointed his own daughter to the U.S. Senate, a staggering example of nepotism."

Comiso warned that, "one of my concerns is if you don't do something, there is always the possibility of abrupt climate change, and there is evidence from the paleo record that this has occurred. Something like that could be catastrophic for humans. Therefore I value doing something about it."

Alaskans "are the canary in the mine shaft," Weller said. He implored people to take notice of climate change and begin to make the needed changes in government, industry and personal activities to reduce greenhouse gas emissions.

Chapter 40

With Our Planet Burning, Mandatory Emissions Reduction Critical

Seeing how massive the oil development around Prudhoe Bay was after mile upon mile on the bus, with scores of miles more to the east, west and south, the scale of the change we have caused is horrific. I had backpacked and rafted scores of miles through the oil fields in 1991, seeing many of the hundreds of oil spills caused every year, along with trash strewn all the way to the Arctic National Wildlife Refuge.

Every well drilled was releasing natural gas that is often just flared or released into the atmosphere. Millions of barrels of oil that was stored carbon for hundreds of millions of years is being pumped out of the ground at an escalating rate. It is like we masterminded the creation of Hell on Earth. Perhaps the Biblical imagery of hell as a fiery place underground was to warn us of climate change. Burning the fossil carbon that was buried under ground is creating hellish conditions that are only worsening.

The year before, I flew from the Arctic Refuge wilderness to Arctic Village, where I spent a few weeks backpacking before flying back to Fairbanks, in a flight I will never forget.

Flying from Arctic Village to Fairbanks, I saw the forests on fire, billowing up thick clouds of smoke from the roaring blazes. There were many fires going, and I could see large areas of blackened earth. The turbulence was intense, shaking the little plane about. As we neared Fairbanks, the smoke was so intense we could not see anything.

In the summer of 2004, fires burned a record 6.7 million acres of forest in Alaska, much of it attributed to the fact the forests had died because of climate change. The fires have continued each year since then. Even way up in the Yukon in Old Crow, a smoky haze limited the view. The Air Force had to close Edwards Air Force Base near Fairbanks. The pilot in our plane noted later that he would have turned back had he known how bad it was, but that he had to land, because we had limited fuel.

Where we could get a glimpse of the ground, vast areas were scorched for mile after mile after mile. We saw several active fires that extended along a line at the edge between the burned area and the partially dead forest. Our planet is burning.

Winds shook the plane as we came in for the landing. I could not see anything but thick gray smoke. I wondered how the pilot could possible make a safe landing. We came down real fast, shaking about. The guy in front of me was so scared that he urinated in his rain boot. After we landed, the guy poured his urine out on the runway, asking out loud if he had violated any regulations. I chuckled, just glad to be alive.

The smoke hurt my lungs. Backpacking around Fairbanks, I knew at once I had to escape. Though wanting to walk about and interview people around Fairbanks, I called up a friend from college who had a cabin with an air purifier near the city and retreated from the lung-damaging smog.

After that flight, I chose to take a bus from Prudhoe Bay to Fairbanks in 2005, also to reduce my carbon footprint and see the haul road. The Haul road is a dangerous gravel road, where oil trucks dominate and everyone else had better beware. The road follows the Trans Alaskan Pipeline, which is raised on stilts the entire distance. The pipeline goes all the way to the port of Valdez, almost 800 miles from Prudhoe Bay.

I was the only person in the van, which had taken a tour group up the previous day. The driver was kind enough to stop frequently at my request for me to get video footage. To see the length to which we have gone to pump out billions of barrels of oil from all ends of our planet, that we have quickly burned is amazing in a very negative way. Oil companies along with their coal industry cronies have worked successfully to stymie action on climate change. First they denied it, then they said it would be nicer on a warmer planet. Now, many admit they were wrong.

Policy makers should have wizened up sooner. Oil companies promised that a major spill was impossible. The Exxon Valdez disaster in 1989 proved them wrong. We cannot trust industry to voluntarily cut emissions. We must mandate emission cuts with a cap and trade system, then rapidly reduce emissions with the mandate of law. The free market can work to find the most cost effective ways to do it. Despite the benefit of the 2009 Stimulus Bill in funding renewables and passenger rail, a cap and trade system is critically needed, along with greater international participation.

Science Writer William K. Stevens called global warming's impacts on Alaska an "ecological holocaust." Christian Science Monitor reporter Ed Hunt found this term fitting, because as in the Holocaust of World War II, so many tried to look the other way and ignore what was happening. Also, as in the earlier part of the second world war, Hunt proclaims, the world's great nations are preparing and uniting to fight, while the U.S. has yet to join in. "I have little doubt we will soon join the fight," Hunt writes. "We can't deny it any longer. Something is happening."

With four percent of the world's population, the United States produces 25 percent of the world's greenhouse gas emissions. Instead of following the guidelines of the Kyoto Protocol, and cutting emissions, the United States is greatly increasing its emissions. They are expected to rise 43 percent by the year 2020 with business as usual.

Passing a cap and trade system to forcibly reduce emissions is the most critically needed legislation in the United States, along with every country on the planet. We can use the principals of a free market to allow industries to compete to meet the demand for energy, wood products and food without generating greenhouse gas emissions. But without the force of law to slash emissions, the crisis will only worsen.

Chapter 41

The Tide is Turning: Climate Skeptics Changing Sides

Pushed by demands for change, many oil industry leaders and their Republican backers are now saying that they realize that mandatory reductions seem to be the logical outcome of this debate, and that they might not be that bad.

In April, 2006, more than 5,000 high school students from more than a hundred communities throughout Alaska signed a petition demanding legislation requiring mandatory emissions reductions. Six students traveled to Washington to meet with Senators and Representatives.

Just after their meeting with Alaskan Senator Lisa Murkowski, Murkowski said in an Alaska Public Radio interview "Oftentimes you need that stick that is hanging out there…I think that may be where we end up at the end of the day." On May 23, 2006, Murkowski voted in favor of Senate Resolution 312 urging U.S. participation in the Kyoto Protocol negotiations.

By 2010, however, Murkowski had returned to her anti-environmental ways, that have been the hallmark of her political career. In January, 2010, she led an effort to gut the power to regulate greenhouse gas emissions from the Environmental Protection Agency.

The tide of realization about the truth of climate change is rising fast. Pat Robertson: said on August 3, 2006: "I tell you stay in doors ladies and gentlemen. Stay cool. Get fans or whatever. And the poor, they need emergency fans and ice to cool down — the number of people dead. I have not been one who believed in the global warming. But I tell you, they are making a convert out of me with these blistering summers. They have broken heat records in a number of cities already this year and broken all-time records and it is getting hotter and the ice caps are melting and there is a buildup of carbon dioxide in the air. We really need to address the burning of fossil fuels." He said that if we are contributing to the destruction of the planet, we need to do something about it.

Former Chief World Bank Economist Sir David Stern predicted dire consequences if we continue the path of inaction on climate change, as he announced a 700-page report commissioned by the British Government in 2006. The report found that the impacts of climate change on the latter half of this century could be on a scale of the world wars and depressions of the first half of the 20th century. "The evidence gathered by the review leads to a simple conclusion: the benefits of strong, early action considerably outweigh the costs."

In May, 2007, global warming skeptic and media mogel Rupert Murdock announced that he was now a believer. He said "It is our responsibility to lead on this issue," and pledged that New Corp will be carbon neutral by 2010.

We have come a long way in the understanding of climate change in the two decades I have been working in my small way to raise awareness about the issue. Carbon neutral was named the 2006 word of the year by the Oxford dictionary. Unfortunately, we should have been this far decades before to forestall the massive damage we have already done. But that means we must be all the more adamant for faster change now.

Having worked to stop or stall climate change initiatives for years, more and more conservatives and industry leaders are now becoming converts to the cause. "This is a creeping disaster," said former Republican Alaskan Senator Ted Stevens. "We're not even sure it's covered by existing (federal) disaster loans." Senator John McCain, chairman of the Senate Commerce Committee, said Stevens' description of the effects of climate change in Alaska are "an argument for doing more than increasing our (computer) modeling capabilities."

Big oil industries are taking interest in climate change, said Climatologist Guntner Weller. "Unfortunately, most are European like BP and Shell. The insurance industry is clearly interested in understanding more and doing more."

Mauer agreed. "Some people in the insurance industry are starting to pay attention, because if weather patterns become more powerful and more erratic and more unpredictable, they might be forking out a heck of a lot more money for damages. Some of the big insurance folks started showing up at climate change conferences as much as ten years ago. So they are starting to wonder if something is going on, and they want to know about it so they can change their calculations on their insurance rates. They can be wiped out if things start changing more than they factor in."

Ernst and Young declared climate change the greatest property casualty threat facing the insurance industry (March 12, 2008). The oldest insurance market in the world, Lloyd's of London, called upon insurers to make global warming the center of their operations. Lloyd's of London Director Rolf Tolle said, "climate change is today's problem, not tomorrow's. If we don't take action now to understand the changing nature of our planet, we will face extinction." (Guardian, June 6, 2006).

Renewable Energy Expert Chris Rose said, "Some of the biggest insurance companies in the world, such as Swiss Re, which is the second biggest insurance company in the world asks its insurers what they are doing about global warming." He said they began doing that "because Swiss Re keeps paying out all of these huge costs for all these extreme weather events. It is the same thing with Munich Re, which is one of the largest re-insurance companies in the world. In the last 40-50 years, the amount of payouts because of extreme weather events has gone up 10 times."

Weller said, "In time I think the popular opinion will swing the politicians around. I think that as long as it doesn't cost too much, there is still the interest in getting an SUV to drive to the local supermarket. I hope that reason will prevail because there is no other way.

"We are a pretty wasteful nation. I think in all levels of society things could be done to improve the situation. Internationally we should talk about Kyoto. On the economic side, I think there are measures that could easily try to achieve the targets of the Kyoto Protocol by fuel efficiency and recycling. I think it is nonsense that it would hurt the U.S. Economy. The Japanese are building fuel-efficient cars. The Dutch and Denmark are building wind generators. There are lots of things that we can do that make money.

"Down at the various levels, city planners are beginning to take it seriously, particularly in coastal areas. You can recycle, you can walk, you can avoid using the elevator, you can drive a more efficient vehicle. There needs to be a moral commitment, and they can actually do something about it. I think the whole engagement of pollution as a moral issue is very important. We need to get the word out that this is something that we should take seriously, and perhaps in our own small way contribute to get people more involved.

Another indication about the solid evidence that global warming is real is that even those who largely caused the problem, and have been trying to discredit the sound science of climate change, are now admitting that it is indeed real.

Even the American Petroleum Institute said, "While consensus on climate change remains a work in progress, we do know enough to take the risk seriously and to rule out inaction as an option."

According to the Intergovernmental Panel on Climate Change, in order to stabilize greenhouse gases in the atmosphere, global emissions must be reduced to at least 60 percent below 1990 levels (80 percent is much safer). That is a radical change in the way the world uses energy. This will require a new industrial revolution with a major decrease in fossil fuels and a blossoming of the already proven efficiency, solar and wind energy technologies. While the American Petroleum Institute is providing lip service to climate change, and maybe a few solar panels on their filling stations, they are not making the real change toward renewable energy and efficiency that is so critical to solving the climate crisis.

Chapter 42

Efficiency: Reducing Climate Change While Helping the Economy

From Fairbanks, I took a bus to Anchorage. The train was tempting, but too expensive. Even in Alaska, mass transit can work in many areas. If it can work there, it can work anywhere. On the bus, I could relax and enjoy the scenery, while meeting other travelers. Driving, I would have been glued to the road for safety, missing most of the scenery and wildlife – except that which jumps in front of my car.

When my brother and his wife took their honeymoon in Alaska, they hit a moose, and my brother was injured with broken glass. It could have been much worse if he had not had his seat all the way back to rest. The picture of their car after the accident was shocking.

Such are the dangers of driving, which kills more than 40,000 people in the U.S. each year, and injures more than half a million. When trains hit moose, the moose meat is given to needy families, and the train is just fine as are the passengers. But it is the energy issue that makes mass transit even more attractive than the safety factor, with climate change trumping all other concerns. After all, climate change threatens the lives of billions of people this century according to top scientists, making even the 40,000+ lives lost each year to auto accidents pale in comparison.

The American economy is among the most energy-wasteful in the world. By increasing efficiency throughout our society from the personal to institutional, we can immediately help to reduce climate change emissions while also improving our economy and making us less sensitive to the costs of energy, which are expected to skyrocket as the dwindling supplies of oil and natural gas are pumped out. Efficiency will also make it far easier to transition all of our energy needs to renewable sources that do not cause any climate change.

Switching back to public transportation can reduce energy by 20 times or more. This has been widely known for decades. The Environmental Handbook, put out by the Friends of the Earth for the first Earth Day, April 22, 1970, reads, "If you wanted to design a transportation system to waste the earth's energy reserves and pollute the air as much as possible, you couldn't do much better than our present system dominated by the automobile.... Compared to a bus, the automobile uses from four to five times as much fuel per passenger mile. Compared to a train, it uses ten times as much. Walking and bicycling, of course, require no fuel at all. Switching from the system of automobilism to a system of rapid transit, with more bicycling and walking in cities, would reduce fossil fuel consumption for transportation by a factor of almost 10."

The same is true today, and even more so with the advent of Sport Utility Vehicles (SUVs). While nearly every other country on earth is getting more efficient with its use of energy, the United States is one of the few that has gotten worse, mainly because of the popularity of gas-guzzling SUVs. Increasing public transportation opportunities and encouraging their use through advertising and subsidized fares is a very needed change.

The United States once had the greatest train system in the world. It was no accident that we changed to using automobiles to the point that now in many parts of the country one cannot travel by public transportation because it is not available. Oil and automobile companies have worked hard to dismantle mass transit, at times buying up rails lines and turning them into roads.

Gale E. Christianson writes in Greenhouse: The 200-year story of Global Warming, published in 1999 of the ruthlessness of General Motors in the 1920s to eliminate rail transportation, and eventually buses. "The lead belonged to General Motors, which was bent on exploiting its advantage by eliminating as many cheap and efficient alternatives to the automobile as possible. Starting in the 1920s, GM formed a series of holding companies that began purchasing dozens of electric surface rail transit systems throughout the country, including those in smog-free southern California, which had the largest interurban network in the land." The holding companies then scrapped the lines, forcing people to take buses or drive. Subsequently, the buses were often scrapped, so people had to drive.

"Patrons of the quiet, high-speed electrical rail service were not used to the noisy, foul-smelling buses and abandoned them by the thousands. The sales of large, fuel-hungry automobiles skyrocketed," Christianson said. And so, we are stuck in the mess of today. We need to switch back, to electrified rail, which is cleaner, faster and far more efficient than driving. We can then electrify the tracks with solar and wind generated energy, thus moving people faster than cars without causing greenhouse gas emissions.

Having traveled tens of thousands of miles on the train throughout the United States, Canada, Europe and Japan, I can attest that it is a wonderful experience, living in the lap of luxury while greatly conserving energy. One looks out a train at the landscape, without the excessive lanes of traffic, berms, guardrails and billboards of the interstate. Instead of the wasted work of driving, the stress and far greater danger, riding in a train one can work, socialize and eat a meal while watching the landscape roll by. It is a quality experience instead of the chore that driving can be.

If an economist could compute the amount of wasted work time that people spend driving, that alone would be a powerful argument to switch back to public transportation. But it is our moral duty to future generations that calls upon us to make the change, and make it fast. We need politicians dedicated to investing in a massive increase in efficient public transportation and a public more eager to take the opportunities of mass transit available while demanding more from government officials.

In the meantime, raising the Corporate Average Fuel Economy (CAFE) standards is in the words of the Sierra Club, "the biggest single policy step the U.S. can take to reduce carbon dioxide emissions." We can save tens of times the amount of oil that could ever come from the Arctic National Wildlife Refuge by raising the CAFE standards to 39 miles per gallon, an improvement easily achievable at great benefit to the U.S. economy. The increase to 35 mpg by 2020 passed in the fall of 2007 is a wonderful step in the right direction, but we must move faster yet. Increasing fuel standards creates many times more jobs through retooling of the auto industry. It greatly boosts our economy by increasing efficiency, while reducing harmful impacts to the planet.

Every other sector of our energy consumption can also be greatly improved. With electricity, switching to compact fluorescents can save 4-5 times the energy of using conventional light bulbs and they last 10 times longer, reducing the energy and resources needed to produce and replace all the incandescent light bulbs that would be saved. LED lighting is even better, and now quite affordable. The same is true throughout our products, and in industrial uses as well.

Energy efficient refrigerators can pay for themselves in electricity savings. More insulation on homes and businesses, turning down the thermostats when no one is in a building (which can be programmed automatically) and a myriad of other, well-known methods can be used to cut our energy use enormously.

For many of us, it is a deep ethic to reduce energy consumption. When one has seen the oil field destruction of pristine wilderness, and the oily toxic mess that inevitably spills, it becomes personal. To see pristine creeks running through forest valleys turn into highwalls and strip mine destruction with acid mine drainage killing the streams and rivers brings the point home that our current energy sources are not right. We should avoid supporting that destruction through wasteful over-consumption. To know that we are contributing to the global catastrophe of climate change should prompt all of us to carefully consider our personal use of fossil fuels, and what we can do to reduce them.

The Environmental Handbook of 1970 states, "Many of us feel that the quality of our lives would be higher with far less use of energy in this country. We would be happy to do with fewer cars, substituting a transportation system that can make us mobile without dependence on the expensive, polluting and dangerous automobile. We would be happy to see the last of the glaring searchlights, neon signs, noisy power mowers and private airplanes, infernally noisy garbage trucks, dune buggies and motorcycles. The quality of our lives is improved by each power plant not constructed near our homes or recreation areas.... Quality of life is a positive ethic. Peace and quiet and fresh air are positive values; noisy smoking machines are negative ones."

Chapter 43

Anchorage Sees Massive Climate Change

The bus reached Anchorage in the evening, with plenty of light left in the northern August day. Fortunately, the bus dropped off right at the hostel I was staying at. It was a crowded hostel, with interesting people from around the world. Many were looking for work, while others were vacationing.

Anchorage has been hard hit by global warming. “The last five or six winters have brought very warm temperatures,” said Alaska Conservation Solutions President Deborah Williams. “I am a cross-country skier. In the last several years there have been very few opportunities for cross country skiing. This is a human impact on winter sports and activities. In Alaska, last year we had to reroute the Iditarod dog sled race because of the absence of snow and frozen rivers.”

“There used to be heavy snowfall in the spring time; there used to be three feet of snow where we walked and now I don’t see it anymore. Instead of dog mushing, we have dog slushing,” said Jimmie Toolie, Savoonga’s eldest elder in a Greenpeace report.

“The latest climate-change anomaly involves the famed Iditarod Trail Sled Dog Race,” The San Antonio Express News editorial reported March 3, 2003. “Because of an abnormally warm winter and little snowfall, race organizers had to ship in snow for the ceremonial start in Anchorage; …they also have had to revise the race route…. In the nation's northernmost state, global warming has become something more than political debate."

"All over Alaska, people are asking the same question: 'where's our winter?'" Rachel D'Oro wrote for the Associated Press December 7, 2002. She reported people were water skiing near Anchorage in December and that trees were budding in the same month.

"Just during the 26 years that I have lived up here, I have seen extraordinary change in snow conditions and winter temperatures and in how Anchorage defines itself," Williams said. "There are fewer days for skiing and snowmobiling. In Anchorage, our main winter festival is called the fur rendezvous. And the highlight of the Fur Rendezvous is the world championship sprint dog races. In the last ten years they had to be cancelled four times because of the absence of snow in February in Anchorage."

The Rendezvous began in 1935, as a winter sports tournament designed to bring people together and bolster community spirit. When it has not been cancelled, it lasts nine days with balls, craft shows, a storytelling competition, snow sculptures, ice bowling, oyster-shucking competitions, the Jim Beam Jam, Cabin Fever Bon Spiel, Pioneer Pancake Feed, Frostbite Footrace, Snowshoe Softball where contestants must drink a beer at each base, auto and sled dog races, canine feats of strength, silly games, time-honored traditions and a carnival. Lori Tobias reported in 2003 that in 1979 when she moved to Anchorage she "remembered days bundled in moon boots, parkas, hats and gloves…. This visit I fly into Anchorage from Portland…I'm still carrying my ski jacket and wearing only a fleece vest over my turtleneck and jeans when I step outside the airport."

The world champion sled dog race is a 25-mile loop around Anchorage at speeds of up to 20 miles per hour. This is in contrast to the 1,122-mile Iditarod race that is run over nine days from Anchorage to Nome with up to 16 dogs per team. The 2003 Fur Rendezvous races had to be canceled because of lack of snow, causing it to be the "furless rendezvous" reported Beth Bragg in the Anchorage Daily News, February 21, 2004.

"A Rondy without dogs is like New Years Eve's without a kiss, the Fourth of July without fireworks, a birthday without a cake…. Two times in the previous three winters, a lack of snow prevented the dogs from running. And both years, Fur Rondy just didn't seem like Fur Rondy."

Bragg quoted Willow Alaska musher Mari Wood's reaction to the cancelled races in 2003. "Every dog musher I know was majorly depressed," she said.

Wasilla musher Sato Konno moved to Anchorage from Japan to race in the 2002 Rondy. A car accident prevented her from racing in 2002. Then, in 2003 the races were cancelled, "a double whammy for Konno" Bragg reported. The Torchlight Ski Parade and family skate also had to be cancelled.

Along with the warmth has come fights and riots at the Fur Rondy. At the 2003 teen dance, fights broke out and police came in with 70 officers, using batons and pepper spray and reportedly threatening teenagers with assault rifles. Teens threw rocks and bottles at police. That was one of the years when many of the winter activities, including the world champion sled dog races, had to be canceled.

"Just a few days ago, ducks floated on little puddles dotting the Delaney Park Strip and umbrellas, not heavy coats, were in demand downtown" reported Zay Hollander in the Anchorage Daily News on February 16, 2003.

Josh Niva reported February 23, 2003, "No snow. No dog sleds. No sled dogs…Warm weather and a lack of snow knocked out the world championship races."

With the lack of snow and unstable ice around the state, trappers brought in only half as many pelts as they had for the previous year's event. Executive director of the rendezvous Mary Pignalberi said, "It is a challenge to pull off a winter festival when you have warm weather." Many dog mushers, snow mobilers and cross country skiers complained about the lack of snow reducing or eliminating their favorite sports. For many of them that was the reason they moved to Alaska.

Nowhere else is the excessive influence of the oil industry more pronounced than in Anchorage, headquarters of some of the biggest culprits of global warming. It is ironic that the Exxon-Mobile dog sled races had to be cancelled because of global warming, as that company is a leader in the global suicidal effort to discredit the sound science of climate change and stall efforts to reduce greenhouse gas emissions such as the Kyoto Protocol.

Williams said, "I really feel that Alaska and Alaskans have a moral obligation to explain to the nation and the world what kind of real, on-the-ground measurable adverse effects that global warming is having. Most Americans still think that global warming is theoretical and they don't really appreciate that it is real and measurable and adverse to the extent that it is, and already being experienced in Alaska.

The day I returned home from Anchorage, hurricane Katrina slammed into New Orleans.

It is not just the polar bear and Inuit who are at risk of the climate crisis. We all share the planet, and we all are at risk. The international organization Christian Aid issued a report in May, 2006 predicting that 185 million people would likely die in Africa this century unless major reductions in greenhouse gas emissions are made.

The 48-page report, "The Climate of Poverty: Facts, Fears and Hope," reported "185 million people in sub-Saharan Africa alone could die of disease directly attributable to climate change by the end of the century."

The organization found that the poor are disproportionately being harmed by the impacts of what the report called the "future shock" of climate change. "That is three times the population of the UK condemned to die because of the spread and increasing intensity of disease, caused by rising temperatures over which they have little or no

Control," the report states. "And that is only the start. What is true for people in sub-Saharan Africa in terms of disease is true for poor people across the developing world."

What's more, the Christian Aid scientists found that those impacts are just the beginning. They found that elsewhere, flooding would displace tens of millions of people, and cause mass starvation with the combination of increased flooding and droughts, as rain becomes increasingly less predictable.

This report, while it did receive some international media coverage, should issue a major wake-up call to people who thought formerly that only the polar bears would suffer. All of those who cannot afford to move from rising sea levels and other climate change impacts will suffer, or die in the not-to-distant future.

Already, Al Gore said that hundreds of thousands of people have died because of climate change. Hurricane Katrina was an example of this in the United States. A half of a degree Celsius temperature rise in the Atlantic Ocean causes an approximate 40 percent increase in hurricane frequency and activity (Nature, January 31, 2008). A 2003 World Health Organization study found that 160,000 people died every year because of climate change. The report predicted this figure to rise massively.

The 2007 IPCC report found climate change will cause between 1.1 and 3.2 billion people to experience water shortages, and up to

600 million people to go hungry by 2080 with business as usual. The report predicted up to 20 percent of the world's population (then expected to exceed 10 billion) will experience flooding, including throughout New York City by 2080.

So it is time to start using accurate language.

Scientist James Lovelock warns that 6.5 billion people could die this century because of climate change. The climate crisis is the number one concern we all should have.

Chapter 44

Cap and Trade Legislation Critical

While the growing reality of the climate crisis is very discouraging, we must not allow the facts to depress our souls. We owe it to future generations to seize this moment and carve out that tunnel of hope through the great mountain of despair, as the Late Reverend Martin Luther King asked those in the civil rights movement during their darkest hours. We must maintain hope that we can tackle the climate crisis, and create a much better, cleaner and healthier world in the process.

The most powerful thing we can do today is get active politically. Join your local environmental organization. Nearly all the major environmental groups are making this a top issue. But don't just stop at paying dues. Get active in making a difference on a daily basis.

Change is happening in America, it just needs to accelerate.

We have the tenacity and ability to do great things if we can harness the national will. With all the resources that we wasted killing people in Iraq and starting a civil war, we could have achieved the targets of the Kyoto Protocol and been on course to greatly reduce our emissions, instead of increasing them as we are today.

Solving the climate change crisis solves so many other problems. Skies will clear of smog, and old growth forests will be restored to sequester carbon (if they can survive climate change). But this is only if we act very quickly. The impacts are only going to get worse and worse, and threaten a future in which forests will not be able to be restored.

Federal legislation to mandate emissions reductions is the single most important step to combating the climate change crisis today. Such legislation would use a cap and trade system to achieve the reductions. This would give a set amount of emissions to industries that produce energy. If they use coal, they will exceed their allotted emissions. They then will need to buy credits from renewable energy companies. These companies would have extra emissions credits because they produced energy without generating greenhouse gas emissions.

This would use the marketplace to find the cheapest way to reduce emissions such as through the use of efficient technology and vehicles, and renewable energy sources. The free market will use American ingenuity to find the best solutions at the lowest cost. The diversity of means to reduce emissions will be far more popular than a top-down mandate that may not suit all people or communities.

The American Council on Renewable Energy published a report in 2007 concluding that renewable energy could supply 50 percent of the nation's current electricity demand and 40 percent of our transportation fuel demand by 2025. The report concluded that there would need to be policy changes to achieve these goals.

Future generations will not be kind to those who procrastinated in these pivotal times. We have precious few years left to vastly alter our course, with permanent damage being wreaked on all future earthlings every day we delay.

While we have known for decades now that we should be lowering climate change-causing pollution, the World Bank found that global emissions have increased by 15 percent between 1992 and 2002, with the U.S. contributing 24 percent of the releases. The rate of increase in the U.S. is about the same as that globally.

We must constantly demand action until our emissions are in fast decline, on pace toward cuts of 80 percent in 30 years, but with actual progress toward that goal every year. We must kick the delayers out of office, and put in champions for combating climate change. Until we can get some solid federal leadership, we must continue action on a state and local level.

Chapter 45

Climate Change Cures Create Better World

While some will bristle at the needed changes, and call them somehow un-American or even communistic, I see things a lot differently. As I picture the world we can create as we reduce greenhouse gas emissions, we can make such a nicer, more aesthetic, cleaner and quieter nation than the concrete jungle of petrol smog that we have today. As more and more people shift toward mass transit we will have fewer roads, highways and parking lots. That space can turn into rail lines, bicycle paths and green space.

Those who enjoy walks through nature will have old growth forest reserves in their backyards as more and more space is given to sequester greenhouse gas emissions. These forests are what every nature lover craves: cathedrals of life with giant trees giving such character and mystical aura lacking in today's second growth. This is hoping that those forests survive the climate crisis.

Food will be grown organically, and locally. This means fresher, better food employing far more people in pleasurable, small scale agriculture rather than the big machines and chemicals needed for the agribusiness of today. Locally grown food needs far fewer fossil fuels to be transported, and thus is one of the myriad of solutions to the climate crisis. A mosaic of organic farms is far nicer than the monoculture fields of large-scale agribusiness today.

The noise and smog of the millions of machines and cars will be nearly eliminated as we switch to solar and wind generated electric vehicles and electric trains. We have become so used to the roar of machines, that we often forget how much it affects our daily life. As I walk in parks and forests throughout the United States and Canada, I am forever plagued by the constant noise of machines.

I cringe at all the noise caused by the internal combustion engine. The sounds of nature are so wondrous, like a grand symphony. But they are easily drowned out by noise. The music of bird songs, waves lapping on the beach, waterfalls, mountain streams, the call of the loon and bugle of an elk is an experience of profound fulfillment. The noise of a car engine in the face of nature's perfection is so horrendously offensive.

But the sound of the train clickety clacking over the rails is music to my ears. Though the whistles can get loud, especially to those living close to the tracks, that is something that could change with overpasses and better signaling methods. In all, I see the changes that must be made as ones that should be made anyway. They will give us more freedom to travel without needing a cumbersome, polluting car.

And they will improve the quality of life for all, in ways we often think we get used to, but never do. All of those noisy cars are jarring our daily senses and drowning out nature's symphony. After walking for hours, in places designated as wilderness, I still hear the roar of cars.

The future carbon-neutral world will be designed so that we can walk or bicycle along tree-lined paths to solar and wind powered trams, taking us to train hubs where we can go anywhere, without the need to own a car, have insurance, and waste all that time driving. Or, we can hop in our small and quiet electric car, charged with the solar panel on the roof, and perhaps some from the local wind farm that charged the grid that we plugged the car into.

Or, just bicycle down to the market along the bike paths created to encourage people to exercise instead of use valuable energy. With saddle bags and carts, enormous loads can be carried using bicycles.

While some may twist these futuristic musings into arguments such as "environmentalists want to push us back to the Stone Age," I say quite the opposite. People could take trains, cars, scooters or battery-assist bicycles charged with solar or wind generated electricity in the future.

What I am saying is that by choosing to bicycle or walk, we can get the exercise that is so sorely needed in America for health reasons, while we do our daily chores. With better-design so it feels like a stroll through a park to go to work or shop, more people will choose to bike or walk to where they need to go, while creating a higher standard of living and reduced medical bills. It is just common sense.

I am not saying that is the only way it is to be done. It will be a combination of new technology, such as solar and wind energy, efficient electric cars and high-speed trains, along with old technologies such as hand grinders, bicycles and trams that will best achieve the needed reductions in greenhouse gas emissions.

By using the exercise we really should be getting for good health to meet our basic needs, we can avoid the need to generate that energy, thus making the conversion of all our energy to solar and wind all the easier. And with efficiency technologies, such as LED or compact-fluorescent lighting we can use far less energy when we do choose to use it.

Using non-toxic natural resins instead of petroleum-based toxic plastics, and recycling existing plastics can help to leave the oil in place where it will nearly permanently sequester carbon, as it has been doing for hundreds of millions of years. Both coal and oil are so dirty and full of toxic pollution that they are best left underground for that reason alone, let alone the need to store their carbon. The most important use of coal and oil is to leave it under the ground, storing carbon to avert the climate disaster that burning it is causing.

Through responsible technology, we can meet our energy needs creatively, by using the methane from landfills and sewage gases to generate power rather than stink up neighborhoods, solar panels on rooftops and wind farms on farm fields, oceans and lakes. This will create clean, safe and sustainable jobs rather than the dirty, dangerous and temporal jobs of the coal and oil industry of today.

The millions who now die or suffer from air pollution today will be spared. The American Lung Association found that there were 24,000 premature deaths caused by the particulate pollution from coal burning power plants in the U.S. every year. And that does not include the mercury, acid rain, ozone, sulfur dioxide, nitrogen dioxide, climate change or the impacts of mining.

The change will take sacrifice and investment. But we must make the transition very fast. We must replace the carbon reserve destroying industries with clean energy used far more efficiently, and do it fast. Through the fast evolution to renewable power, we can create a much better quality of life for everyone. If we care about the future, we really have no choice.

End

About the Author

Born in Columbus, Ohio, Kister backpacked 700 miles from the oil fields of Prudhoe Bay into and throughout the Arctic National Wildlife Refuge in 1991, and published the book Arctic Quest: Odyssey Through a Threatened Wilderness about the journey. Kister has personally lobbied numerous U.S. Representatives and Senators to protect the refuge over the last 15 years. He has studied, written and spoken about climate change for nearly 20 years. In 2000, Kister was a climate campaigner with Ozone Action (now part of Greenpeace USA), and was a delegate to the United Nations Conference on Climate Change in the Hague, Netherlands in 2000.

For the last seven years, Kister has traveled more than 100,000 miles throughout the United States and Canada doing more than 1,000 multi-media presentations. Kister has traveled mostly by fuel-efficient train. In 2005, Kister joined the Board of Directors of the newly formed Arctic Voices organization, and was appointed vice-chair of the non-profit. Arctic Voices will strive to show the human face of the Arctic Refuge issue through the voices of the Gwich'in people.

Kister returned to the Arctic in 2005 with a video camera, and produced his first film, Caribou People, available at www.arcticrefuge.org. Mostly in the words of the Gwich'in and Inupiat peoples, the film has amazing wildlife footage and shows the threat that oil development presents to both cultures. Gwich'in say development would cause a genocide of their culture, and scientists concur that it would have a massive negative effect on the Porcupine Caribou Herd that the Gwich'in depend upon.

Kister lives in Athens County, Ohio where he works locally in the progressive community to change to solar and wind energy production, mass transit and bicycle paths and lanes, locally produced organic food and sustainable, energy efficient housing. The author/filmmaker also works to protect public lands in the area from destruction by the rampant coal mining, logging, oil and gas development and unnecessary highway construction on national forest and other public lands.

The author will continue to tour the continent doing presentations about the Arctic Refuge, Climate Change and solutions to both. In 2007, he published Against All Odds: The Struggle to Save The Ridges about the so-far successful effort to protect 700 acres in Athens, Ohio that are owned by Ohio University, which had originally proposed to develop Radar Hill. The land had long been a community gathering area. After an intense struggle, Radar Hill was declared a nature preserve. The rest of the land, however, needs permanent protection.

This book can help to inspire people to get active politically, showing that even when it seems hopeless, with enough community involvement and perseverance you can prevail against all odds. It is an exciting narrative with lessons for any nonviolent struggle anywhere in the world.

Upcoming books by Kister include:

- a book about Kister's amazing journeys through the Tongass National Forest and southeast Alaska: the largest old-growth forest in the United States. It is being massively clearcut.
- a sequel to "Arctic Quest" about his journey with a group from Old Crow into the Arctic National Wildlife Refuge in 2004.

Color photos and maps of Kister's 700-mile journey through the Arctic are at www.arcticrefuge.org where a CD with high quality maps of the entire Arctic Refuge and hundreds of photos can be ordered. "Arctic Quest" and "Arctic Melting" are also available on audio at www.chadkister.com.

He can be reached at www.arcticrefuge.org, www.chadksiter.com, chadkister@gmail.com, (740) 707-4110.